Industry Response to Health Risk

THE CONFERENCE BOARD

Report No. 811

ABOUT THE CONFERENCE BOARD

The Conference Board is an independent, not-for-profit research institution with facilities in the United States, Canada and Europe. Its scientific studies of management and economics produce a continuing flow of timely and practical information to assist leaders of business, government, labor and other institutions in arriving at sound decisions. The Board's research is also made available to the news media in order to contribute to public understanding of economic and management issues in market economies.

Worldwide, The Conference Board is supported financially by more than 4,000 Associates, comprised of corporations, national and regional governments, labor unions, universities, associations, public libraries and individuals.

THE BOARD'S SERVICES

Research reports, personalized information services and access to a variety of meetings are among the direct benefits that Associates receive from The Conference Board. Associate Relations representatives at the address listed below will describe these activities in detail and will tailor delivery of Board services to the specific needs of Associate organizations. Inquiries from locations other than in Canada and Europe should be addressed to The Conference Board, Inc. in New York.

The Conference Board, Inc.
845 Third Avenue, New York, New York 10022
(212) 759-0900
Telex: 234465 and 237282

The Conference Board of Canada*
Suite 100, 25 McArthur Road, Ottawa, Ontario K1L-6R3
(613) 746-1261

*A cooperating institution operating under license.

The Conference Board, Inc.
Avenue Louise, 207 - Bte 5, B-1050 Brussels, Belgium
(02) 640 62 40
Telex: 63635

Conference Board Report No. 811   Printed in U.S.A.

©1981 THE CONFERENCE BOARD, INC.

ISBN No.:0-8237-0248-0

Industry Response to Health Risk

by Audrey Freedman

A Research Report from The Conference Board

Contents

S.O. 12.8.81

APPENDIXES

Exhibits

About This Report

THIS REPORT is based upon interviews with executives and corporate employee relations, environmental and health staff at companies with extensive "leading edge" experience in developing health and environmental response; interviews with health, union and industry associations; mail and telephone inquiries to other experts on individual points of information. Because the elements of corporate response described in this report are relatively new, and companies are dealing with uncharted territory in public health programming, an extensive bibliography is provided.

Foreword

SINCE 1970, an increasingly comprehensive net of federal and state laws has imposed public health responsibilities on private industry. At its essence, the recent legislation asks for more than compliance. It requires industry to detect and engineer its way out of health-damaging processes and products.

This study explores the possibilities of a private industry initiative to produce public health. A careful examination suggests that, for the most part, the public authority must always be the genesis of public health guardianship—because market incentives do not exist to activitate enterprise. On the other hand, some companies help the government with information. They provide some of the identification of hazard; they provide in-

formation on how to control the hazard. If such company help were not provided, the public authority would be, more and more, sucked into detailed operation of private enterprise.

The Conference Board thanks the company executives who provided information and insight during extensive interviews. Also, we thank those professional association, industry association, union and academic specialists who contributed specific information. The study was carried out in the Public Affairs department, Walter A. Hamilton, Vice President.

KENNETH A. RANDALL
President

Part I

INTRODUCTION

Major Public Health Legislation in the United States

Direct, legislated imposition of public health responsibility has increased in scope and coverage, especially during the 1970's. For example:

1906 Pure Food and Drug Act
1938 Food, Drug and Cosmetic Act
1947 Federal Insecticide, Fungicide and Rodenticide Act
1966 Federal Hazardous Substances Act
1967 Wholesome Meat Act
1968 Wholesome Poultry Products Act

1970 Occupational Safety and Health Act (OSHA)
1970 Clean Air Act
1970 Hazardous Materials Transportation Act
1972 Ports and Waterways Safety Act
1972 Consumer Product Safety Act
1974 Safe Drinking Water Act
1976 Toxic Substances Control Act (TSCA or TOSCA)
1976 Resource Conservation and Recovery Act (RCRA)

Chapter 1
Public Health: Industry's New Responsibility?

FOR THE LAST DECADE public concern over the influence of corporations on public health and employee health and safety has reached new highs. Not surprisingly, this area has become a prime target of regulations and a fertile area for litigants.

Essentially there are two approaches—the first is economic; the second legal. An economic understanding of the situation facing private firms is based upon the production of a "public good" (such as education, or defense). The issue has been: How will this society bring about the production of the "public good" of health? A legal or legislative understanding moves to the next stage and emphasizes the sanctions that influence private-firm behavior. A third stage has begun to appear in large, sophisticated firms—a management response and approach. This report details the results and difficulties of these early efforts to organize and manage dual production of a marketable product and public health.

The Economic Viewpoint

It is an axiom that the primary purpose of a business in the private enterprise system is to produce goods and services for profit. The choice of *what* goods and services will be produced is largely left to free market forces.

However, there may be some socially desired "goods" that private industry does not find it profitable to produce—goods that are not really marketable to individual purchasers. A "public good"—such as decreased risk of lung cancer through less pollution—cannot be sold to individuals because it would be available to all members of society. When private producers cannot sell the benefit or good, they cannot recover their costs—and will not voluntarily produce it. Society uses government to compel the production of a "public good"—in this case, public health.

The failure of a market system to generate public health is only half of the reason for the advent of regulatory compulsion. The other half is the undesirable effects, external to the corporation, that a company's activities can impose on society. These spillover, external costs are public costs. The objective of some regulation, and a good deal of civil litigation, is to force the *internalization* of such costs. Theoretically, if a company knows that it will have to pay the costs for all of the illness and disability it causes, it will devote itself to avoiding (or producing less) disease.

In justifying regulatory imposition of public health responsibility on private industry, the spillover effects of production activity have been most often cited. For example, the air quality and lung disease effects of coke oven operation are external effects. The social costs of operating a steel mill—incuding early disability and death—may not be borne by private companies *through the market system*. That is, the selling price of steel may not include the full cost of treating those diseases that its production engendered. In economic analysis, this situation is called a "market failure."

One way of bringing social costs inside the private company is by taxing firms to compensate the victims. This is the essence of the coal mine "black lung" compensation system. Lawsuits such as product liability actions may operate like a tax, also, in charging back to the firm the spillover effect of its activities.[1] These kinds of taxing systems appear to be easier to devise when the social cost is imposed by a clear-cut event (such as an accident) and measurable costs (such as loss of an arm).

The effect of regulation is to force the external costs back inside: For example, the insulation producer is required to pay for previous damage—the cost of curative medical treatment; the cost of dependency. This

---

[1]However, in acting as a vehicle for remedying market failure, individual lawsuits create their own social costs. These are principally the administrative expense of case-by-case fact finding and adjudication. See E. Patrick McGuire, *The Product-Safety Function: Organization and Operations*. The Conference Board, Report No. 754, 1979.

internalization of cost, where it can be accomplished, is an indirect form of regulation relying primarily upon the price system. For free market advocates, it is a preferable form of assuring the production of public goods.

The second alternative is *directly* to require firms to operate in a less health-threatening manner. In effect, this route is to impose upon firms a required level of public health production.[2] It is in this sense that the health legislation listed on page viii has imposed upon some private firms the obligation to produce a "public good."

A Lawyer's Approach

The use of government to cause private industry to internalize the social costs of production is an economic description. Legislators and lawyers, describing the role of government, might begin with the sanctions—as in this account.

"Those individuals and organizations found liable for civil or criminal penalties under one or more toxic pollutant laws ... face potentially stiff monetary sanctions and, in criminal actions, the possibility of prison terms up to two years. Beyond this, however, other sanctions may be experienced. For example, the company involved may be barred from eligibility for federal procurement.... Essential operating permits may be revoked, causing substantial economic loss. Under TSCA [Toxic Substances Control Act], manufacturers of a substance found to pose an unreasonable risk to the public may be ordered to replace or repurchase it. In addition, of course, there may be penalties under state law and claims for damages under common law or state statutes, including derivative suits against responsible officers. Finally, the greatest 'penalty' of all for a toxic pollutant violation may well be the disapproval of the public, reflected in reduced sales to consumers and a tarnished company reputation...."[3]

Entering from a lawyer's point of view raises the terminology of "compliance"—a somewhat different imagery from that of producing a new "public good" product. Yet, the legal explanation arrives at many of the same private firm actions and internal programs that are described later in this report:

"The positive corporate compliance program ... must flow from a clearly communicated attitude on the part of top management that compliance *is* essential, that candor in surfacing potential problems is desired, and that those who take prompt corrective actions—which often require significant expenditures that are non-productive for purposes of the immediate profit-loss statement—will be rewarded rather than penalized. Unless such an attitude is clearly conveyed to subordinates by top management's behavior as well as by memos, it is doubtful whether subordinates will effectively implement any compliance program.

"Beyond attitude, of course, there must be a management structure capable of putting the proper attitude into practice with effective results. Normally, in any large business enterprise, especially one whose operations are geographically dispersed, there should be a central environmental manager and staff to coordinate plant-by-plant activities, to assimilate and communicate applicable rules and policies, and to provide a check against the frequent tendency of operating officials to ignore or avoid environmental problems. Whatever the exact structure, however, there must be clear assignments of responsibility and procedures to ensure accountability. These must include an explicit mechanism for bringing unpleasant problems—e.g., malfunctioning control equipment or disturbing toxicological tests on a product—to the surface for remedial action at a responsible level.

"A major hurdle that often must be overcome in a company is the instinctive attitude that 'the less I know about my emissions, etc., the better.' This sort of attitude often produces decisions to do only that amount of monitoring of discharges or testing of products that is absolutely necessary to meet government requirements. Purposeful ignorance is almost always self-defeating. If the added information from more adequate testing is favorable, it can only strengthen your position should a government or citizen group challenge arise. If, on the other hand, the added information is clearly negative, its early discovery, prior to an enforcement action or other lawsuit against the company, maximizes your options for solving the problem, may permit avoidance of legal sanctions, and may well save substantial costs. Indeed, the cost of monitoring and testing, compared to the costs of penalties or more stringent regulations, is usually an excellent investment."

---

[2]In economic terms, the private firm is also required by this alternative to bear the social cost of production and this higher cost will be reflected in its prices.

[3]Roger Strelow, "Corporate and Individual Responsibilities and Some Suggestions on Preventive Law," in *Toxic Control,* Volume IV, Proceedings of the 4th Toxic Control Conference, Washington, D.C., Government Institutes, Inc. December 10-11, 1979, pp. 170-176.

The Corporate Approach

The reactions of major private firms to this situation is no doubt changing and will continue to change. Knowledge improves and successes emerge. Companies of lesser size and economic impact than the pioneers will implement these organizational, procedural and technical changes to build in public and employee health protection during the manufacturing, processing, or handling of their products.

Some aspects of industry's response resemble traditional new-product developments. Other aspects are heavily flavored with the "defense" tactics of response to government intrusion. This report focuses its attention on the "new product" aspects of the corporate response—though all response to external pressures has defensive elements.

One of the major themes of corporate response is selling the new concern internally, because these new products are not a profit-seeking response to market opportunity. This internal sales job often comes under such compliance rubrics as "informing operating management about the impact of new regulation." In a way, such a description suggests that companies only respond to outright, specific regulatory orders. This is possibly true—but another possibility also emerged during the company visits on which this report is based: Corporations develop specific internal advocates and change agents in response to new public demands. These agents may cite government rules to amplify their voices inside the corporation. In this sense, they benefit from the pressure of government regulators and plaintiffs' lawyers. But their role is a positive one—to be an internal generator of change.

The process of internal change has only barely begun. As various health-related laws were passed, corporate applications and refinements needed to be worked through. The response *within* many corporations is still only at initial stages. Every company studied described its systems as changing and fluid. Many, commenting on a particular control, say: "We just introduced this last Summer"; or "Next January, we are going to reorganize that structure"; or "This is too risky now—a lot more training will have to be done first."

As with new-product development, there is a good deal of technical expertise required to design and engineer the public health "product." Moreover, changes in the content and goals of some technical fields are being required. This kind of shift is not only internal to the corporation as an entity, but internal to the individual employee whose job is modified. The engineer now considers additional specifications. The financial officer must add a new perspective. The plant operations manager must incorporate new information on costs and inputs into the long-range operating agenda. The epidemiologist has a role in private corporations for the first time.

The fluidity of these responses marks this report as one frame of a moving picture. The chapters that follow describe *some* of the responses (and issues and problems) of some very large and sophisticated corporations that are aware of health issues. It is possible that the change in the Administration, and lessening of regulatory pressures, will create a pause. Yet the public pressure for accountability is less changeable than Administrations.

In a recent survey conducted for Marsh and McLennon by Louis Harris and Associates, different groups were asked questions about many varieties of risk. With regard to occupational disease, all groups focused major responsibility on the corporations. The question was: "For a disease caused by exposure on the job to dangerous substances years earlier, who should have the primary financial responsibility?" The answers are tabulated below.[4]

Responding Group (number of respondents)	The Employer	The Employee	Shared Responsibility	Not Sure
Top corporate executives (402)	70%	2%	22%	6%
Investors and lenders (104)	74	3	19	4
Public (1,488) [a]	83	5	10	3
Federal regulators (47)	85	2	11	2
Congress (47)	89	—	9	2

[a]Percentages do not add to 100 because of rounding.

[4]Marsh and McLennan (New York), "Risk in a Complex Society: a Marsh and McLennan Public Opinion Survey," 1980, Table III-7. The survey dates were December, 1979 through March, 1980.

Part II
How Companies Organize to Manage Health Responsibility

Chapter 2
People and Things

THREE THEMES underlie company moves to institute organized health-control systems. The first can best be described as *initiative,* or the spontaneous development of health concerns and the subsequent dedication of corporate resources to health stewardship. Parts of the company systems described in this chapter date back to such initiatives, often taken several decades ago. The second theme can be characterized as a *response to government regulation.* As regulations imposing health responsibility on private industry have increased, particularly during the late 1970's, companies have moved to institutionalize and control their operating response: to organize for compliance. The third theme is *reaction to calamity,* the discovery that a major health hazard has been found in a company process or has occurred because of a company's product.

The company systems that have developed by the early 1980's are probably a product of all three themes interacting within a specific company context. The companies studied do not describe their systems in terms of having been "required by government regulations." Nor are health and environmental efforts organized entirely under the rubric of "compliance." For example, some elements of health compliance systems predate OSHA (Occupational Safety and Health Act), or at least predate specific OSHA regulations—such as those dealing with exposure monitoring. Moreover, the systems are usually described as having grown out of different efforts at different times and as being "unique to our company."

On the other hand, the adversarial pressures of public opinion, muckraking investigations, lawsuits and regulatory intrusion—especially when they converge on an industry simultaneously—are powerful stimulators. Some would say that the *anticipation* of such adversary developments can be an incentive for corporations to assume preventive public-health responsibilities. For example, a company may fear that its actions could provide fuel for more regulation, or that it will suffer in the marketplace. A company may be concerned to protect its corporate reputation, or that of the industry in general. Such motives might be interpreted as a response to market pressure, rather than to government intrusion.

One way of categorizing company health-protection systems is to differentiate between those that focus mainly on people (primarily employees, but also customers, customers' employees, those who live near company facilities, those who come in contact with company products or wastes), and those that deal with things (substances, processes, equipment). Essentially, this division draws attention to the two specialties that predominate in company health systems: medicine and engineering. In nearly all company interviews, these two fields were represented when the company selected the appropriate people to discuss its health programs. These are the pivotal fields in leading-edge companies, although other departments—law, employee relations, and public relations—may also be involved.

The engineering-medical staffing suggests that companies are engaged in an active response to public health concerns. (A passive or defensive response might be to staff the departments primarily with lawyers and public relations people.) If industry were totally passive it would treat health responsibility entirely as an imposition from outside. It would await detailed instruction from governmental authority, focusing much of its effort on resistance. The government, in order to ensure compliance with OSHA, the Resource Conservation and Recovery Act (RCRA), and other laws, would become a partner in monitoring the health of an industry's employees, testing each company's processes and operations, engineering each company facility. In such a situation, nearly all moves are initiated by regulators external to the company and industry. In such instances the company stance could be: "You tell us exactly what to do, how to do it, and, if the courts ultimately say we must, we will follow the letter of the requirement."

Some government regulation does follow this pattern. For example, the minimum wage law (Fair Labor

Standards Act), is imposed with detailed government specifications—by industry and even by employer—as well as the intrusion of government investigators to check payrolls. Yet, the minimum-wage legislation is a relatively simple requirement compared with any one of the health laws listed above (page viii). Some would argue that legislated health responsibility could never be imposed in detail, and enforced by intrusion, because it is so complex.[1] So the regulators are dependent on regulatees

for information, for design of specifications, and, ultimately, for compliance itself.

Another possible reason for the medical and engineering approach in corporate compliance efforts involves the possibility that the health laws are effectively requiring industry to incorporate public-health needs as by-products of the industry's regular output (see Chapter 1). If this is the case, then companies might elect to deploy designers and production and testing forces to develop and begin producing this "new product." In this instance the product designers and producers (as well as the internal marketers) are medical, engineering, toxicological and epidemiological scientists. Perceived this way, corporate health-compliance programs would have the configuration shown in Exhibit 1.

---

[1]Industry does, however, complain of overly detailed regulation in the "worker safety" area. See James Greene, *Regulatory Problems and Regulatory Reform: the Perceptions of Business.* The Conference Board, Report No. 769, 1980, pp. 10-11.

Exhibit 1: Some Inputs to the "New Product"

People; medical theme	Products and processes; engineering theme
Health surveillance systems:	Redesign of equipment and/or job function to affect exposure
• employee physical exams	Redesign of product to alter ingredients or process of manufacture
• exposure records	Introduction of protective equipment
• location-of-work records	Introduction of monitoring equipment
• occupation records	
Studies and analyses of employee:	Added (or new) weight given to potential health *impact* of waste and waste disposal, of possible uses of company product, of transport and transport malfunction, of "surprise" in product itself, and so on.
• health facility visits	
• health benefit (diagnosis) claims	
• workers compensation claims	
• "outside data" (Social Security, etc.)	

Chapter 3
Health Surveillance Systems

THE FOUNDATION of company health compliance systems is a surveillance, or monitoring, function. Surveillance systems are constructed from sets of information on employees and on company processes. First, they may be designed to *originate* new facts by investigating and recording something new to company interest. An example would be periodic employee health evaluations. Other information banks may be developed out of facts already of interest and "known," but not organized for analytic purposes—for example, what chemicals the company uses (or produces), by specific location, or by location and occupation of the individual assigned.

The creation of any one data file may require a major company effort demanding, for example, the institution of periodic physical examinations, tailored to a preestablished protocol, for various segments of the work force. This alone constitutes a health surveillance system of the first order. But, combining these data with employee exposure data (themselves a product of data files covering the location of work and materials used) produces a surveillance system for more complete analysis. One way of perceiving these subsystem components, and their part in an overall surveillance scheme, is shown in Exhibit 2. It depicts Diamond Shamrock Corporation's software package COHESS (Computerized Occupational Health/Environmental Surveillance System).

Surveillance systems that integrate employee health and exposure data permit epidemiological investigations within the work force. That is, the company can use its system to spot higher-than-normal incidences of disease or symptoms (e.g., reported neurological symptoms) and detect whether the incidence is correlated with a particular pattern of exposure, job function, or locale of work. Going in the other direction, a system can be used to find out if the group of workers exposed to a particular substance, or working in a particular process or locale, have higher-than-normal pathological changes.

Such studies, based as they are on statistical patterns and "normal" probabilities, require large employee populations. Therefore, as long as surveillance systems are designed and implemented within single companies (or divisions of companies), they are only useful to major employers.

The cost of surveillance design and implementation also operates to confine the initiative for establishing such systems to large and profitable companies. In the chemical industry, which has a natural interest in occupational health, the Chemical Manufacturers Association identifies nine companies—all major producers—as having fully implemented health surveillance systems.[1]

As noted, smaller companies find that the most obvious deterrent to building a health surveillance system is cost. However, it may be useful to ask: What if cost were not a factor? Then, the technical issue stands out: There are insufficient numbers of employees in small companies for the detection of patterns. For example, if only a half-dozen of the company's employees are exposed to a particular substance, and at three different locations where three different configurations of companion substances are present—a search for statistical pattern is not feasible.

The need for large numbers of employees, either to detect adverse health effects—or to *disprove* them—suggests a collective system. If only a few employees per company are in contact with a substance, then an effective health surveillance detection scheme would necessarily require collective, central computerization methods. Creation of such cross-company files would not only require substantial (and mutual) resource commitments from competitors in industry, but also a level of collaboration and information-sharing that is unknown at the present time. A hospital consultant's report on the

[1]Interview with Dr. Geraldine V. Cox, Chemical Manufacturers Association, Washington, D.C., July 14, 1980.

Exhibit 2: Diamond Shamrock's Computerized Occupational Health-Environmental Surveillance System (COHESS)

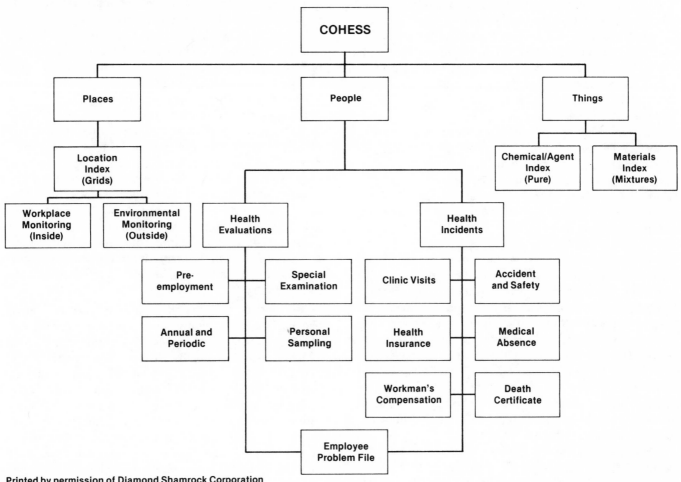

Printed by permission of Diamond Shamrock Corporation

business opportunity for health-care providers observes: "The medical surveillance market is totally driven by OSHA regulations based on TOSCA (Toxic Substances Control Act) recommendations. In talking with large corporations (10,000 + employees), we found the data protocol and computer facilities were internal to the organization. However, the existence of the six external software providers may be an indication of how firms employing less than 10,000 are receiving their data service."

The foregoing discussion appears to assume that health surveillance systems are strictly defined—not only as to purpose and use, but also in terms of specific components. If this were true, it would be possible to count them: which companies "have them," which do not. However, The Conference Board interviews show that computerization is perhaps the only concrete similarity that appears in all discussions. There are individual employee health data—but these vary from company to company or by such factors as occupation; locale; division; presumed exposure; and so on. There are data

on exposure, but they vary greatly in the original source of fact for entry, in the time periods covered—prospective and retrospective—and periodicity of entries. There are also wide differences in opinion about the feasibility of including data that are derived outside of the firm's developing surveillance system (e.g., diagnosis items from the employee's health benefit claims).

Variation also appears in the purpose or the *use* to which health surveillance systems will be put. Epidemiological use—actively searching for abnormal patterns to identify possible disease-causing exposures—is not universally foreseen. Some companies have a "monitoring" (defensive) purpose in mind; that is, to ensure that employees as a group do not exhibit unusually high incidence of morbidity or mortality.

The following description points up some of the interactive effect of purpose and design on such systems:

"Medical surveillance examinations must be planned and designed with the known hazards of the particular group of employees in mind. The earliest effects of any hazardous exposure, just like the earliest effects of many diseases, may be subtle and require special tests. Therefore, the medical surveillance examination must be a skillful blending of (1) general tests to assess general health and to make sure there are no great gaps of as yet unrecognized hazards, and (2) appropriately selected special tests to detect early effects of specific exposures which an employee is known to have had." (Excerpt from an internal information manual supplied by a manufacturing company.)

That company is using its system to generate the appropriate medical test protocols. It associates individual employees with physical work locations. Operating staffs enter the substances used at those locations, and industrial hygienists enter the monitored amounts of substances present (e.g., in the air). According to company memoranda:

"These three inputs define the employees who may have been exposed to possibly harmful influences, and therefore, who should have a medical surveillance examination to search for early signs of damage. The system will not only set forth what medical surveillance examination should be performed because of present exposures, but will also indicate what examinations should be performed because of past exposures. Knowledge of past exposures is important due to the latency period of carcinogens. Additionally, ongoing medical surveillance will be required to guide follow-up actions as new carcinogens are identified.

"Because of time constraints on the employee, all these various exposures, past and present, need to be synthesized into an ongoing, current exposure profile of an employee which changes depending on new assignments and new work sites. From this composite, individual exposure profile, a single multipronged medical surveillance examination can and must be 'tailored' for that employee since medical examinations are preferably given on an annual basis only."

Surveillance Systems as Research Tools

The ultimate use of health surveillance systems may be epidemiological analysis of change in health patterns in the work force. In theory, a sophisticated, computerized health scan could search a corporate—or an industry—work force, seeking out unusually "high" patterns of morbidity and mortality associated with a location, a particular process, or a chemical.

In fact, at the present time, employees themselves—through their unions—are posing epidemiological issues in two prominently discussed cases. The United Auto Workers union (UAW), on behalf of its members, began a few years ago to question whether there was an unusually high rate of cancer among wood model shop employees. In November, 1979, the union announced that it was asking the National Institute of Occupational Safety and Health (NIOSH) and the Michigan State Department of Health to do studies, and that there would be a joint UAW-General Motors study of 15 model shop locations and work forces.[2] The president of the Michigan Cancer Foundation, which conducted a preliminary study, commented: "There is an incidence passing beyond chance, and that has to be faced up to. To do that, you have to dig in and find out."[3]

In another auto industry situation, the United Rubber Workers asked NIOSH to determine the cause of skin disease patterns among employees using a "wet rubber process" at a Dayton, Ohio plant.[4] NIOSH, in turn, asked General Motors for the medical records of employees at the plant. This incident shows that union members, functioning as shirt-sleeve epidemiologists, may initiate a study by a government agency. As one union medical consultant commented: "For sheer manpower, the workers represent the greatest potential contribution to an industrial medical surveillance system in the country."

Investigation of the incidence of cancer among chemical and petroleum workers has been growing, with major studies reported in 1980. For example, a study of 17,000 people employed by a Texas refinery between 1935 and 1979 did not show increased risk of brain tumors. Another study, among employees of eight other refineries, showed no increased incidence of brain tumors, but excesses of esophageal, stomach, intestinal, rectal and nasal cancers, and melanoma.[5] One of the largest studies under way covers the health and work histories of 40,000 Union Carbide employees to discover whether certain cancers are especially prevalent in the employee group. The company is working with the University of Pittsburgh's School of Public Health (which did a preliminary study of workers' death certificates) and with NIOSH. A company spokesperson commented to the news media: "It's going to be quite a detective job. This is . . . the first time there's been any major joint effort between government and industry to conduct a study like this."[6]

There are two approaches to identifying relationships between disease and exposure. One, referred to as "case control," concentrates on a group of individuals with the

[2]UAW press release, November 9, 1979.

[3]*The New York Times,* February 14, 1980.

[4]*5 Daily Labor Report E-1—E-2.,* January 8, 1981.

[5]From an unsigned article entitled, "In 1981: Brain cancer probe to intensify in chemical and petroleum industries," *Occupational Hazards,* December, 1980, pp. 57-58.

[6]UPI, "Union Carbide Study To Relate Chemicals and Employee Deaths," August 17, 1979.

Exhibit 3: Du Pont's Case Control Study Approach

Case Finding

In epidemiological studies, every person in the study population who develops the disease under investigation must be identified and included. Otherwise the incidence and prevalence of the disease will be underestimated and measures of the relation of risk factors to the development of the disease may be rendered inaccurate.

The problem of case finding can be especially difficult in a company such as Du Pont because of its large size and the widespread dispersion of its employees throughout the country. Where the onset of the disease under investigation generally results in death or prolonged disability, such as heart disease, cancer, and stroke, complete case finding has been accomplished by means of claims filed under two company-sponsored insurance plans. One is a group accident and health insurance plan for nonoccupational illnesses or injuries which cause disability of eight days or longer, and the other is a life insurance plan that covers both active employees and those who have retired on a Company pension.

Where the onset of the disease frequently does not result in disability or death, such as diabetes mellitus, alcoholism, and hypertension, we have relied for case finding on medical records and surveys of Company physicians.

Study Designs

We have made use of retrospective, cross-sectional, and prospective study designs, either alone or in combination, depending on the availability of information and the objectives of the study. To investigate factors that influence the prognosis of persons who have developed a particular disease, we have conducted long-term follow-up studies of cohorts, one consisting of a series of persons who have the disease and another consisting of a control group matched to the persons with the disease.

A case-control design has been incorporated into almost all of our studies. For each person who develops the disease under investigation, a control is selected at random so that the case and control are matched with respect to age, sex, geographical location, and socioeconomic status, as indicated by whether the employee is a production worker or is salaried.

Measuring the Degree of Excess Risk

The risk that a person with a particular characteristic will develop a certain disease is measured by the incidence rate: that is, the number of persons with the characteristic who develop the disease over a specified time period divided by the total number of persons with the characteristic in the study population.

The relative risk is the ratio of the incidence rate among persons with the characteristic to the incidence rate among persons without the characteristic. For example, if the incidence rate of coronary heart disease among persons with hypertension is 6 per 1,000 persons per year and the rate among normotensives is 3 per 1,000, the relative risk is 2.

For some of the characteristics included in our studies, it was not feasible to get incidence rates because the total number of persons in the Company with the characteristics (the denominator of the rate) was not known. We were able, however, to compute estimates of relative risk by applying certain statistical techniques to data obtained from cases and controls.

Source: Sidney Pell, "The Identification of Risk Factors in Employed Populations" from *Transactions of the New York Academy of Sciences (series II, vol. 36),* April, 1974, pp. 342-3.

same disease or pathological symptom (e.g., high blood pressure). Occupational histories, exposure data, other health histories, "life-style" items—all elements which might have contributed to the condition—are sought and introduced into the analysis. (See Exhibit 3.) With sufficient information, it is theoretically possible to find out how much one of these elements contributed to or increased the risk of disease. Also, some conclusions may be reached on the interactions of several factors. One researcher points out: "When two or more chemical pollutants act jointly on an organism, the real possibilities arise that their toxic actions may simply be *additive,* that one may *amplify* the action of another and vice versa . . . or that one may *block* the action of another (italics added).'" Finally, it may be discovered that a single substance can cause multiple effects.

[7]Seymour L. Friess, "Contribution of Statistics to the Analysis of Environmental Health Problems Caused by Pollutants." *The American Statistician,* February, 1977, volume 31, number 1, p. 3.

Many such retrospective studies have two characteristics: They are based on current cases of a particular disease and in a particular work force (e.g., rubber-tire workers). Sometimes, the particular work force will be defined more broadly—for example, those who are exposed to lead. Possibly the group will be selected geographically: those living and working around a petroleum refinery, steel plant, or nuclear power installation. The key element is potential exposure.

Starting with presumed exposure, and directing a study toward the detection of resultant disease, can require as much reconstruction as starting with the disease. For example, a brief description of a study of the effects of dioxin exposure in one plant notes that the medical team, directed by Dr. Irving J. Selikoff of Mt. Sinai School of Medicine, "conducted physical examinations including X-rays, biopsies, biochemical blood tests, and compiled medical and occupational histories. Present and former employees of the plant were examined."[8] Such a study of exposed individuals is designed to detect an excessive incidence of disease or of abnormalities (e.g., blood cell abnormalities), or of such other "effects" as birth defects or low birth weight among workers' children. (See box.)

The study of exposed groups can yield negative findings as well: for example, a finding that there is no higher incidence of abnormality in the exposed group than in other groups matched as to working status, age, and other characteristics. Sometimes, a negative finding leads to a refinement of the original hypothesis. The following situation illustrates this back-and-forth search mode in process.

The Chemical Workers union asked OSHA and EPA (the Environmental Protection Agency) to ban a particular herbicide because it was alleged to cause birth defects among children of workers producing it. The union's evidence—gathered from one plant—convinced those agencies and NIOSH to undertake studies. NIOSH is surveying personnel records, ascertaining exposure to the herbicide and other chemicals in the plant. A survey of reproductive problems encountered by workers during a six-year period is being made, and the union's evidence is being checked against hospital records.

EPA has surveyed eight *other* plants that currently or previously produced the herbicide—and found no unusual levels of birth defects. The NIOSH study director, taking this negative result into account, commented: "At this juncture, we know that the number of congenital defects among the offspring of workers there (at the union-cited plant) is clearly excessive. But what's causing them is something else. Is it dioxin, or some other chemical, or is it just bad luck? The answer to that

---

Research Opportunity Noted in 1974

Through its periodic examination and other health programs, industry is a repository of a vast amount of uncorrelated and unexamined information on health history, demography, absenteeism, occupational exposure, and other factors related to disease. A subject that is sometimes discussed in occupational medicine circles is how industry's unique ability to generate such health data for large and diverse populations over a long period of time can be put to proper research use. A uniform or comparable data system—toward which the OSHA reporting system may be a first step—would, it has been suggested, create a source of inestimable value toward a better understanding of occupational influences on health.[1]

[1]Lusterman, Seymour, *Industry Roles in Health Care.* The Conference Board, Report No. 610, 1974, p. 28.

---

question is going to require a lot more digging.'"[9] "Bad luck" may mean "undiscovered causal agent" after more intensive study, but this kind of result is increasingly unacceptable to the union.

Attempts to reconstruct exposed groups for study purposes often reveal the paucity of appropriate records. For diseases such as cancer, work exposure 20 to 30 years previously may have been the triggering event. The existence of decades-old payroll records with names, addresses and social security numbers, may be "a matter of good luck" (and that is possible only if the employing company is still in business). Finding the employees now is difficult and expensive. Even the Federal Government's resources, such as social security records, are poor tracers: They were not designed for this sort of task. Companies that had life insurance and death-benefit plans for production workers, or had pension plans that vested employees with some coverage, may have some records that can be used to trace employees who were not on the payroll at retirement. The easiest to "find" are those employees in a stable company, plant and process who spent a lifetime in one job, retired and are collecting their monthly pensions.[10]

Even under the most favorable record-keeping circumstances, complete reconstruction of an exposed group has not always been possible. In 1977, Du Pont's Medical Director, discussing a firm's duty to report health hazard, observed:

---

[8]Peter J. Sheridan, "What's causing mysterious illnesses?-NIOSH seeks answers," *Occupational Hazards,* April 1980, p. 66.

[9]Sheridan, p. 69.

[10]It is possible to reason that this remnant of a cohort exposed, say, 25 years ago, might have a greater likelihood of showing disease because its exposure may well have lasted longer.

"There is also considerable discussion at present on how to inform former employees who may have worked with newly identified chronic health hazards. This concern has been generated in large part by discovery by the National Institute of Occupational Safety and Health (NIOSH) that it may have some responsibility in this area. Unfortunately, we cannot offer any easy solutions ourselves. This problem, quite frankly, reveals gaps in our reporting system, gaps which we feel must be closed. We are still trying to find a good method to locate employees who left the company perhaps as long ago as 20 and 30 years, but who may have worked with a material now suspected of being carcinogenic.

"Some of this information is available to us through the epidemiologic studies we have been doing for more than 20 years. We are also working to establish employee work and exposure histories, but we have a long way to go in these areas. The goal of informing all these different groups of people is still largely elusive, although in some of our retrospective epidemiologic surveys we have managed to reach 98 to 99% of the identified cohorts."[11]

In some cases, often because a material was already suspected of being dangerous, companies began keeping exposure records years ago. One of the managers interviewed for this study described these records:

"We have never really done anything with them. We never put them on the computer. Well, computers in this field are very new—only being used in the last five years. Nobody used computers back in the 1930's and 1940's. There is no way we can sit there with ten guys with green eyeshades on spitting on their fingers leafing through pages. It took a while until we finally figured out what we wanted, and then we had to present this to top management to get a budget to do it. But we didn't go back in time. But it isn't useful until you can use it as an early warning system. And, you can't function that way until you have had an accumulation of a lot of years of experience. You have to put in whatever you have going backwards. We have several epidemiology studies under way right now—massive studies."

Reconstruction of an exposed group for epidemiology will include finding those who died, as well as the survivors. Death rates by age will be of interest, but the search will also, of course, include the cause of death. One company described its search this way:

"We have to go to Social Security to find out if they are alive or not. If we have the last-known address, then each one of these people must be called; we must try to

talk to some near relative. Then we must make a judgment as to how reliable that person's answers are. We then try to get the death certificate and find out the cause of death. Death certificates are really not that reliable; they may say one thing, and mean something else. This company employs a specialist in reading and interpreting death certificates: a nosologist."

Company epidemiologists say that it is critical in retrospective studies to recreate as much health and exposure detail as possible. This is to guard against attributing a high level of disease solely to the *known* exposure. For example, if the diseased study group also had a higher-than-average level of cigarette smoking—and some of the excess disease was due to this factor—such an attribution or discounting cannot be made without the basic raw material: data on each individual's smoking habits.

An illustration of this point occurs in criticism of a NIOSH study of effects of beryllium exposure. *Science* reports:

"Bayliss had also charged that insufficient attempt was made to determine if the lung cancer victims were smokers. The industry had claimed, on the basis of sketchy interviews with the victims' relatives or friends, that as many as 30 may have smoked. The article discounts the potential influence of smoking on the survey results by noting that many of the tumors found in the 47 workers were adenocarcinomas, in lieu of more common smoking-induced bronchogenic carcinomas.

"Infante responds to Bayliss' criticism that 'as is known to anybody familiar with the acquisition of such data, retrospective ascertainment of cigarette smoking habits is extremely unreliable'—a difficulty compounded by the fact that many of the victims' spouses are also deceased. Carl Shy, an epidemiologist at the University of North Carolina who served on the HEW review panel, adds that an absence of information about smoking is common to retrospective epidemiological studies, including some of those linking cancer with exposure to uranium, asbestos, and acrylonitrile. Yet the scientific community has accepted evidence for an association between lung cancer and these hazards, Shy says."[12]

Other criticisms of the beryllium study also illustrate how "missing items" may bias the results. The researcher claimed that an insufficient attempt was made to find out *how much* exposure to beryllium the cancer victims had experienced. He also suggested that, if more recent data

[11]Bruce W. Karrh, "A Company's Duty to Report Health Hazards," *Bulletin of the New York Academy of Medicine,* September, 1978, pp. 782-788.

[12]R. Jeffrey Smith, "Beryllium Report Disputed by Listed Author." *Science,* volume 211, 6, February, 1981, pp. 556-557. © 1981 by the American Association for the Advancement of Science. (David Bayliss is an epidemiologist, who worked on the beryllium study in question while he was an employee of the National Institute for Occupational Safety and Health.)

on the incidence of lung cancer had been used, the "expected cases" among exposed workers would have been higher. Then, the "actual cases" observed would not, in his view, have been a statistically significant excess.

Surveillance Systems as Monitors

Health surveillance is used in some companies to monitor exposure reactions. For example, direct monitoring of atmospheric dispersion is supplemented, in a sense, by indirect monitoring for detectable health effect. By searching the potentially exposed employee groups for subclinical changes or signs, before clinical symptoms develop, the company has an early warning of trouble.

Screening for signs of trouble may draw together several separate sources of data. (See Exhibits 4, 5 and 6.) A preemployment physical contributes baseline measures for the individual. It is used for comparison with later measurements to detect change. Data on individual occupational exposure are added. From this latter source, the appropriate tests are generated—the company medical department knows "what to watch." A third item to be included might be reported visits to the company health unit and the nature of the "complaint" or diagnosis.

One of these worker-monitoring programs, operating at Burlington Industries, illustrates in specific detail how the "systems" approach is crucial to the usefulness of the program. Dr. Harold R. Imbus, corporate director of health and safety, described the program to *Occupational Hazards* this way:

"The major ingredients of the Burlington medical screening program for cotton dust are the spirometer exam and the British Research Council Respiratory Questionnaire Modified for Byssinosis . . . . The information obtained identifies such things as chronic cough with sputum production, Monday chest tightness, shortness of breath, smoking habits, and work history including exposure to cotton dust. At each Burlington cotton processing plant, the staff nurse, using this questionnaire, interviews workers.

"Pulmonary function testing on a spirometer is conducted on the first day of the new work week, before a worker begins his shift. We use a spirometer . . . . The instrument measures how much air is expelled in .1 second (FEV.) and also measures the total amount of air pushed out of the lung (FVC.) . . . ."

The other monitoring system elements are direct:

"Several types of corporate audits are conducted during the year. To double-check reported levels of cotton dust within the plants, Imbus' industrial hygiene group travels to plants in a special truck filled with sampling devices and equipment. Air samples from each plant area are forced through filters, which are weighed before and after the audit in Burlington's Greensboro labs. Fiber levels from the filters are measured and compared with the OSHA standard levels: 200 micrograms of lint-free respirable cotton dust per cubic meter of air, averaged over 8 hours for yarn manufacturing processes; 500 micrograms per cubic meter for non-textile industries; and 750 micrograms per cubic meter for slashing and weaving operations.

"Another corporate audit involves an engineer-and-nurse team who visit the plants and check all OSHA

Exhibit 4: Excerpt from a Conglomerate Company's Internal Report on Occupational and Environmental Protection Activities in 1979

THE OCCUPATIONAL ENVIRONMENTAL AND HEALTH INFORMATION SYSTEM

After conceptual planning and a feasibility study, approval was secured to proceed with the development of this information system consisting of three parts:

(1) A toxic substance control information system (TSCIS) module, including a chemicals data base and other health and safety data, will provide capability for meeting reporting and record-keeping requirements of TSCA, and produce material safety data sheets. Programming for this module is presently underway and is scheduled to be operational later in 1980.

(2) The industrial hygiene module will record workplace exposures to hazardous substances and physical agents to meet record keeping and reporting required by various regulations, and facilitate epidemiological studies.

(3) The medical record system is already established and in the process of modification and enhancement by the medical group.

These modules will be interactive and provide a multicompany data gathering, updating, searching and reporting capability that should prove uniquely comprehensive and cost effective.

Exhibit 5: Uses for Monitoring Data for Research and Evaluation[1]

"At the present time, the data generated by medical and environmental monitoring are used primarily for protecting the health of individual workers. However, the data are a potential source of information for epidemiologic studies and for evaluating the effectiveness of OSHA's health standards.

"There are, however, several major problems with using OSHA medical and environmental data for these purposes.

—Only active workers are subject to medical surveillance. For example, in the case of chronic disease, especially cancers with latency periods of 20-40 years between exposure and disease onset, workers may have left the workplace or retired before the disease became evident. If there are high turnover rates in a hazardous industry, many at-risk workers will be hard to locate from follow-up studies.

—Only exposed workers must be offered screening. To get the most accurate estimate of disease risk and to determine the effects of differing exposure levels, data should be collected on all workers in a firm. This would permit internal comparison of the health status of workers with little or no exposure to that of workers with higher levels of exposure.

—Some OSHA standards require medical monitoring only of people with exposures above the Action Level which in most cases has been one-half the permissible exposure limit. This limits the range of doses for which health outcome data can be obtained.

—For epidemiologic studies, consistency in what data are collected and how they are collected (methodology of sampling and analysis, types of instruments, quality assurance etc.) is necessary to permit pooling or comparison of results from different firms. This means greater consistency among the standards than currently exists and more specificity for the types of tests to be done and the methods of sampling and analysis to be used. Recent standards on lead and cotton dust provide this specificity.

—At the present time, most companies have no mechanism (such as a unique identifying number) to permit easy linkage of medical and work history records, job descriptions, exposure data and medical test results. This is especially important since OSHA standards do not consistently require that medical surveillance records contain this essential information. This omission gravely limits the use of data in establishing the relationship between exposures and disease, and the magnitude of that association in relation to different exposure levels.

"While OSHA's medical and environmental monitoring data are of limited use for research and evaluation purposes, these data can potentially provide some short-term or proxy measures of program effectiveness.

—Assuming the reliability of environmental monitoring data, and assuming that compliance with an OSHA standard is equivalent to some degree of health protection, environmental data for a given industry or a sample of firms can be looked at over time to detect trends in airborne levels of toxic substances.

—For some diseases, medical screening tests will indicate an adverse effect on a worker's health fairly rapidly after the onset of exposure. For example, workers exposed to certain substances like asbestos and silica will show an excessive annual loss of pulmonary function. Measurements of blood lead levels are fairly sensitive indicators for evaluating changes in worksite exposures. In such cases, medical screening data on groups of workers could be analyzed over time to see if improvements have occurred since the standards were issued.

"Given the limited usefulness of OSHA's medical and environmental data for purposes of research and evaluation on the effectiveness of health standards, it is necessary to consider the use of longitudinal cohort studies. The value of cohort studies of large groups of workers is illustrated by such studies of asbestos and textile workers. Because of the careful studies of large groups of such exposed workers, a good deal of scientific data is available on these substances. In addition, NIOSH is currently conducting a longitudinal cohort study of coal miners to evaluate the effectiveness of the Mine Safety and Health program.

"Additional analysis is needed to examine the feasibility of systematically establishing and funding longitudinal studies of workers in a wide range of industries, as well as alternative ways for constructing such cohorts of workers. It appears that such studies would overcome the major problems associated with the use of employer medical and environmental monitoring data. For example, a longitudinal cohort analysis would: (1) include retired workers and better account for job mobility, (2) provide for an adequate control group to account for intervening or confounding variables by systematically selecting and testing exposed and nonexposed workers, (3) assure uniform data collection and analysis and permit linkages between medical and work history data, exposure data, demographic and other data, and (4) include multiple exposures to hazardous substances."

[1]Excerpt from Draft Report of the U.S. Dept. of Labor, "An Interim Report to the Congress on Occupational Diseases," December, 1979, pp. 119-122.

Exhibit 6: Input for Medical Surveillance (From a manufacturing company training slide)

Responsible Department	Input
Operations Industrial Hygiene	Designate plant work locations; identify potentially harmful substances present
Employee Relations	Census; and general employee demographic information; job title ("track" employee through plant work locations)
Medical, Toxicology, Epidemiology Industrial Hygiene, Operations	Determine which substances are possibly toxic to humans at generally used levels, and use this information to set internal company standards
Industrial Hygiene	Determine actual levels of potential toxins and environmental hazards present in the workplace
Toxicology	Inform; which body organs are hit by the toxin?
Medical, Epidemiology	Design an examination to detect early evidence of effects or damage from the toxin and/or environmental hazard
Medical	Design epidemiological studies of employee populations to assess health status compared to general population.

The Company identifies a group of employees exposed to a toxin(s) at a level greater than is encountered in the group normal environment, who are examined using procedures designed to detect early signs of damage to the organs particularly attacked.

Epidemiology Review
Feed Back
↓
Operations, Engineering, Industrial Hygiene

Clinical Review
Feed Back
↓
Employee

compliance procedures referring to work practices, maintenance, medical programs, and recordkeeping. Imbus noted that the audit, initially conducted twice a year, is now due on an irregular, unannounced basis."[13]

Since Burlington began its program years before the OSHA cotton-dust standard, it has been able to reduce symptomatic "Monday morning tightness" from 4.5 percent of tested employees in 1971 to .6 percent in 1979.

The adverse physical effects of overexposure to some materials can be halted and, in some cases, reversed. For example, this is true for cotton-dust exposure. In some cases, in time the body can rid itself of certain amounts of a chemical. (But, the chemical may also undergo changes, and the resulting metabolites "target" upon an organ to which they are harmful.) Health surveillance systems that are designed for monitoring can warn management to intervene. As the Director of the Environmental Hygiene and Toxicology Department of Olin Corporation commented at a scientific symposium: "The ultimate decision of whether a potential hazard is being adequately controlled is determined by careful periodic medical evaluation of each individual."[14]

[13]From an unsigned article entitled, "Medical monitoring and surveillance in action." *Occupational Hazards,* September, 1980, pp. 43-44.

[14]Richard Henderson, "Thresholds for Control of Potential Hazards in Occupational Environments." *Journal of the Washington Academy of Sciences,* June, 1974, p. 134.

Chapter 4
Informing Employees of Their Physical Condition

THERE WERE clear-cut differences of viewpoint among companies visited on informing employees about their own physical condition. (See Exhibit 7.) In some companies, the medical department will discuss an employee's condition, if asked, but will provide no specific data and will not allow the employee to see his or her own medical file.

One manager described this practice in policy terms: "We do not believe that the employee should sit down with his or her own records and look at them and leaf through them. This is simply because the average production worker is not capable of reviewing such records. The person doesn't know what the medical terms mean. Our doctors have been instructed for many years that they must sit down with the employees and they have to tell the employees what is wrong with them."

This position is accompanied in some companies by the offer to share information on the employee with the employee's family doctor in direct doctor-to-doctor communcation. Then, the family doctor may explain the employee's physical condition to the person. Company medical directors, in particular, feel that employees are too unsophisticated to be able to understand their own health measurements and records, and that giving such data to an individual could cause distress. On the other hand, it was a company medical director who referred to these policies as "the medical ethics of the 1930's" in which patients knew nothing; doctors were all-knowing and shared knowledge only within their own profession. Nonetheless, a survey of the *Fortune* 500 companies, by Professor David Linowes of the University of Illinois, reported in *Occupational Hazards* in December, 1979, showed that 83 percent of responding companies do not allow workers to see their own medical records.

Some companies expressed concern that, if an employee were given his or her own record, it might be turned over to the union. Then, perhaps, the union would develop the possibility of "outsider" health surveillance. Both medical directors and labor relations departments were concerned about this possibility in companies where unions have raised major occupational health issues.

The Occupational Health and Safety Administration issued a regulatory standard on employee access to medical and exposure records, effective August 21, 1980. The regulation covers employees "exposed to toxic substances or harmful physical agents." Under the rule, an employee can see and copy his or her own medical examination and test results (e.g., lead level in the blood), medical history, diagnoses, treatments, prescriptions and medical complaints. The employee may also find out to what materials he or she is being exposed (or was exposed in the past, if a record exists), via Material Safety Data Sheets. With the employee's consent, his or her representative (the union) may also see and copy the medical and the exposure data and analyses based on these records. Exposure records and analyses must be maintained, under the 1980 regulation, for 30 years; and medical records for the duration of employment plus 30 years.

A law firm newsletter identified several "problem areas." One is that the employee may ask the union "to obtain other exposure records and conduct a study to see if the incidence of any particular diseases are increased in the exposed population." Also, an employee may claim workers' compensation for an illness based on exposure; or may try to sue the company doctor for malpractice. The law firm suggests: "The prudent employer would do well to have an employee [company doctor or nurse] translate rather than have the [requesting] employee take the records to a friend or a lawyer for interpretation." The law firm's comment concludes:

"As a result of some of these problem areas, employers may be deterred from providing full-scale medical services to employees because of the extra problems that care could generate. As well, *employers may approach with great caution requests to become involved in epidemiological studies or conduct their own studies of substances* to which employees may be exposed because the right of employee access extends to all research based 'at least in part on information collected from individual employee exposure or medical records.' "[1]

In similar vein, a Wharton School study warns:

"An employer must decide whether to go on collecting as much and as detailed information as had been the case before. An industrial hygiene program's output will now have to be seen as a potentially public output from the company to employees and to designated nonemployee recipients. Reexamination of what is collected and why and in what detail will be essential to the employer's successful coexistence with the new rule."[2]

Access to information was the subject that elicited the most heated opinions during Conference Board visits. Company medical directors referred to medical ethics and professional standards as reasons for not giving employees information. Professional traditions were cited: Doctors share information "only with a licensed physician." However, the subject seems to be undergoing change and some deeper examination of motivation. For example, Dr. Lloyd B. Tepper, medical director of Air Products and Chemicals, Inc., said in an address to a medical conference:

". . . It was simply not the general practice 30 or 40 or more years ago to disseminate information to those who might be affected by materials or processes. . . . It was the style of the time for those in authority to do what they thought was right without full participation in or understanding of the basis of the decision on the part of those affected. Needless to say, this style of establishment paternalism is largely moribund if not dead . . . . There is a right to know, and in large measure the duty to inform lies with us in the health professions."[3]

The OSHA standards on access had just become final at the time many of the interviews were conducted for this study. It seemed likely that company physicians as a group were in the process of reexamining their institutional viewpoint on access by the patient or employee. Possibly, the OSHA standards speeded and deepened this reassessment.

[1]Seyfarth, Shaw, Fairweather, and Geraldson, *SSF & G Labor and Legislative Report,* Chicago, July, 1980, pp. 7-8.

[2]James T. O'Reilly, *Unions' Rights to Company Information.* Labor Relations and Public Policy Series No. 21, Philadelphia, Industrial Research Unit, The Wharton School, The University of Pennsylvania, 1980, p. 138, reproduced by permission.

[3]From an unsigned article entitled, "OSHA's new medical disclosure standard: Fundamental right or threat to privacy?" *Occupational Hazards,* August, 1980, p. 49.

The code of ethics adopted in 1976 by occupational physicians (see Appendix C) covers the subject at a level of generality that omits mention of individual workers' access to information. It suggests that the physician should:

". . .treat as confidential whatever is learned about individuals served, releasing information only when required by law or by over-riding public health considerations, or to other physicians at the request of the individual according to traditional medical ethical practice; communicate understandably to those they serve any significant observations about their health."

The inclusion of "when required by law," however, seems now to have given the rule a specific content with regard to *employee* access, because the OSHA regulation is law.

Access by outside investigators, particularly those at the National Institute of Occupational Safety and Health (NIOSH), has been fought in the courts primarily on the ethical issue of employee privacy. Two such cases, decided late in 1980, have settled some issues in favor of disclosure. In one case, the Westinghouse Electric Corporation raised employee privacy concerns as a reason for not providing health information to NIOSH for epidemiological studies—unless each employee affirmatively provided written consent. The records contained a report of the initial hiring physical (including chest X-ray, pulmonary function, hearing and visual tests, blood count, and medical history) and reports of subsequent physicals for exposed workers. NIOSH had performed its own tests, and wanted earlier data for comparison to detect deterioration, if any. The Third Circuit opinion was that "the strong public interest in facilitating the research and investigations of NIOSH justify this minimal intrusion into the privacy which surrounds the employees' medical records, and that Westinghouse is not justified in its blanket refusal to give NIOSH access to them or to condition their disclosure on compliance with its unilaterally imposed terms."[4]

In a similar appellate court case, the Sixth Circuit ruled that NIOSH may subpoena the medical records of employees to carry out an occupational disease investigation. In this case, the Institute had asked General Motors for the records of workers using a "wet rubber" insulating process. General Motors had invoked the employee's "right to privacy," and an Ohio statute concerning "the confidentiality of physician-patient communications." The court responded:

"We reject this analysis by noting that the Ohio privilege statute is not the controlling principle of law here. This case presents a federal question; the applicability of a privilege must, accordingly, be ascertained

[4]Quoted in 211 *Daily Labor Report,* A-9, October 29, 1980.

Exhibit 7: Allied Corporation's Policy on Access to Medical Records

The following company policy statement covers access to medical and exposure information by employees, company staffs, and government agencies.

A. Internal to the Company

1) *Employee Access*

Employees, former employees, or their designated representatives shall have the right to inspect and copy, all or in part, his or her medical record. All such requests must be honored within a reasonable time and within 15 days of the initial request.

With respect to medical information whose disclosure to the employee may, in the opinion of the Plant or Location Physician, *have an adverse impact upon the health of the employee,* such as a specific diagnosis of a terminal illness or a psychiatric condition, the Company may inform the employee that access will only be provided to a designated representative of the employee having specific written consent. *Where a designated representative with specific written consent requests access to such information, the Company must provide access to the designated representative, even when it is known that the designated representative will give the information to the employee.* However, the physician, nurse, or other responsible health care personnel may delete from the requested medical records the identity of a family member, personal friend, or fellow employee who has provided confidential information concerning an employee's health status. To the extent practicable, inspection of the record should be made in the presence of an Allied Chemical Physician, Company Medical Director, or Nurse who shall endeavor to explain the meaning of the content of the record to the employee and *recommend,* as appropriate, that the employee:

a) Accept a summary of material facts and opinions in lieu of the records requested, or

b) Accept release of the requested records only to a physician or other qualified medical professional.

Note: Rebuttal of information contained in the medical record by the employee will be considered:

• If the proposed statement is acceptable to the Allied Chemical physician and is signed and dated by the employee, it will be placed in the medical record.
• If the rebuttal statement is not acceptable to the Allied Chemical physician, a note must be made in the record that a decision to refuse to accept the rebuttal was made.

2) *Allied Chemical Occupational Health Professionals Access*

Access to the medical record on a "need to know" basis by hygienists, epidemiologists, and/or toxicologists for purposes of research and statistical studies must be authorized by the responsible company medical director. These individuals are bound to the same confidentiality requirements as the physician and nurse.

3) *Allied Chemical Safety Professionals Access*

Allied safety personnel shall receive appropriate summaries of occupational illness and injury incidents and shall have access to plant dispensary and/or first-aid logs in keeping with their responsibilities for providing a safe workplace.

4) *Other Allied Chemical Access*

It is understood that pertinent medical records or data can be released, all or in part, to the corporation's insurance carrier, legal counsel, employee relations representatives, or other authorized representative in connection with disability and workmen's compensation claims or other similar actual or pending claims against the corporation. Confidentiality is waived in such instances. *However, Allied Chemical medical services personnel, wherever possible, shall take special care to see that only information relevant to the claim is disclosed.*

B. External to the Company

General Rule

As a general rule, and subject to the exceptions specified below, Allied Chemical will not release to external sources medical information concerning an employee or former employee unless the individual has authorized such release by providing the Company a signed and dated Consent for Release of Medical Information, Form No. C-3551, or its equivalent. Consent must be provided for each instance of release and will be honored only if presented to the Company within 60 days of the date of authorization by the employee.

Specific Rules

1) Routine Requests for release of Medical Information

With the written consent of the employee, a written request by a physician, medical institution, other authorized health agency or insurance company, for abstracts or copies of part or all of the individual Medical Record will be honored. Allied Chemical's specific "Consent for Release of Medical Information," Form No. C-3551, or its equivalent must be signed.

2) *Medical Emergencies*

In the event of a medical emergency involving life threatening circumstances whereby information contained in the Employee Medical Record is deemed important to the immediate care of the individual, information contained in the record should be released by the Allied Chemical Medical Department upon the request of a responsible family member or the attending physician.

3) *Release of Pertinent Medical Data to Appropriate Public Health Authorities*

When it is determined that a public health issue or risk has been uncovered, as in the case of reportable communicable disease, appropriate notification to state or municipal health authorities shall be made.

4) *Access by Government Agencies*

a) To preserve the confidentiality of Employee Medical Records, it is the policy of Allied Chemical to request government agencies to seek release of such records only upon the written consent or authorization of the employee. However, the Company policy notwithstanding, OSHA has the authority to require immediate access (OSH Act, Section 1910.20, Access to Employee Exposure and Medical Records, Revised and effective August 21, 1980) to employee and former employee medical information and/or analyses that do not contain personally identifiable information. Whenever OSHA seeks access to personally identifiable medical records or analyses of current and former employees *without the prior written consent of the individual* it must present to the Company a *Written Access Order* in accord with their rules of agency practice and procedure governing OSHA access to individual medical records. Once received, all such Written Access Orders, including a copy of the Order and its accompanying cover letter, must be prominently posted for at least fifteen (15) working days.

*IMPORTANT NOTE: *Before any Medical Records are released to a government agency without prior written employee consent, the approval of Company, Corporate Medical Services, Corporate Employee Relations, and the Corporate Law Department should be obtained in order to determine whether such agency request falls within the regulatory authority of the agency. This procedure can be facilitated by contacting your Company Medical Director or Corporate Medical Services immediately upon any request for access.*

5) *Access by Designated Representative*

The Company, upon presentation of a written consent by a current or former employee, Form No. C-3551 or its equivalent, will release individually identifiable medical records to a designated representative. "Designated Representative, means any individual or organization to whom an employee gives written authorization to exercise a right of access." With respect to medical information that may be deemed by the Company Physician to have a detrimental impact upon the health of an employee, the Company will be governed by the procedure outlined in Section VI, Access to the Medical Record, Part A, 1). Also, information received in confidence from sources outside the Company that is included as part of the Medical Record *will not be* divulged to the employee's designee.

Special Note: Consents for release of medical information will be honored only if presented within 60 days of the date of authorization.

6) *Access to Analyses, Epidemiological, and Corporate Statistical Studies*

Allied Chemical will release upon request Company initiated, composite statistical data regarding occupational health matters to employees, designated representatives, and OSHA, so long as such studies *do not* contain individual and personal identifiers regarding employee medical records. All such requests for release of this type of data *require* the prior approval of Company, Corporate Medical Services, and Corporate Law Department. *Employee consent is not required in such instances.*

7) *Access by Job Applicants to Written Preplacement Examination Results*

Job applicants *do not* have the right to inspect or receive a copy of the preplacement physical examination or the written physician's recommendations other than by subpoena.

8) *Access by Subpoena*

Properly issued subpoenas for the access to medical records will be honored. The responsible manager at the location at which the subpoena is served shall notify his Company Management, including the Company Medical Director, and the Corporate Law Department, in accord with Allied Chemical Policy 122.0. It is also required that the Director of Medical Services, Corporate Headquarters, be immediately notified.

Special Note: Allied Chemical reserves the right to charge a reasonable fee for expenses involved in photocopying of records.

9) *Employee Information Posting Requirements*

Allied Chemical must notify an employee, upon employment, and all employees annually thereafter of the existence, location, and availability of medical records and analyses thereof covered by the OSHA regulation, 1910.20. Allied Chemical must also make available to employees a copy of the Standard, 1910.20, and must distribute to employees any informational materials concerning the referenced regulation which is made available to the Company by OSHA. A copy of the recommended record accessibility notice for posting on plant bulletin boards is attached, see page 14.

10) *Company Information to Employees*

The Company has available for training and communication to employees a Video Cassette Training and Information program including Video Tape, Leader's Guide, and Employee Brochure describing Allied's philosophy and approach in dealing with employee medical records. Copies may be obtained through your Company Medical Director.

---

by reference to federal statutes and the common law . . . . The common law did not recognize a physician-patient privilege at all . . . . Neither has Congress codified the concept in a federal statute. A decision in this case based on considerations of the physician-patient relationship would, in effect, expand the scope of the 'federal common law.' This we decline to do."[5]

With access to company medical files slowly being opened for public health purposes, what may unfold next? One area of skirmish may be the diagnosis reports that are included on application forms for health benefits. These are often filed by employees or doctors

with the company's benefits office, then passed on to the insuror for payment. As long as the company sees these reports, however briefly and for whatever purpose, it may "be obliged to perform the affirmative act of collecting that data for review by an employee or employee representative."[6]

In an effort to analyze and gain some insight and control over the costs of health-care policies, employers have become more interested in doing their own claims administration. But claims administration means temporary possession of diagnosis. So, a counterincentive may develop out of occupational health pressures—it may seem better for an employer to have nothing to do with health-insurance claims. Yet, when the need to trace the incidence of disease patterns becomes strong enough, NIOSH may turn directly to the insurors' records of diagnosis and treatment.

[5]Opinion in General Motors Corp. v. Director of the National Institute for Occupational Safety and Health, Department of Health, Education, and Welfare, decided and filed December 30, 1980, reprinted in 5 *Daily Labor Report*, E-1—E-2, January 8, 1981.

[6]*SSF and G Labor and Legislative Report*, July, 1980, p. 6.

Chapter 5
Monitoring the Company Environment: Industrial Hygiene

CORPORATE CONTROL of hazards in the workplace environment is centered on industrial hygiene. Essentially, this is the measurement, to a standard, of the airborne, liquid and direct-contact substances that can be dangerous to humans. The basic ingredients of an industrial hygiene policy are covered in Exhibit 8.

Industrial hygiene standards for substances (either as input materials, intermediate substances, by-products, wastes or final products) are established internally— often at the corporate level. Then, on-site monitoring observes reality against the standard. Such monitoring can be continuous (e.g., for stack gas emissions) and may be engineered to start an alarm-alert process in immediately dangerous situations. On the other hand, it may be as infrequent as once a year, when company policy has specified this interval for an "audit."

The Vice President for Health, Safety and Environment of a worldwide minerals company commented on his regular (scheduled) industrial hygiene audit program this way:

"An industrial hygienist goes into one of our operations to monitor for dust, fume, vapor, noise and any additional environmental conditions that they think should be monitored for. They issue reports to the people in this department as well as the operating people. If there is nothing wrong with the operation, if it meets all federal requirements, if it meets *our own* corporate standards (and they can be different because our corporate standards may be more stringent than the government standards), there is no need for further action on the report.

"If for some reason or another they do not meet these, and there are many reasons why they would not meet them, then action is taken immediately to correct the situation. If it is a condition perhaps where the bag house, for example, has not been maintained properly, before the hygienist left the plant he would say to the plant manager: 'Your system X Y Z is not performing up to what we think it should be. Please look into it.' So in most cases, by the time the official report has been issued, the plant manager has taken steps to correct the situation. Sometimes the situation is corrected in a matter of days, and sometimes it is corrected in a matter of hours.

"If it is a more serious problem the plant could shut down. For example, I was informed by phone at 10 o'clock one morning that levels of solvent were just outrageous in one plant being audited. Within half an hour the line was shut down. It was not allowed to start up again until they had taken corrective measures. In this case, the corrections were very time-consuming, so the line was shut down from six to eight weeks. When this happens once, word spreads and they make mighty sure that it doesn't happen again."

When the measurement results of monitoring show an unacceptable situation, several courses of action may be initiated. At the plant level, work practices and "housekeeping" may be tightened up, and the industrial hygienist may become diagnostician and consultant to local management in getting the job done. Or, some minor engineering or equipment changes may be necessary; local management may develop these with its own staff or some outside help. A variant of this was described in one interview at a minerals mining and processing company:

"The plant manager is put on notice when there are marginal readings, even though they are below the government TLV (Threshold Limit Value.) If they are one-and-a-half now, they could be two-and-a-half tomorrow. We don't want to see that. So, he starts to work to get these refinements. The hygienist will often help by passing the word: 'So and so has tried this and it's working. Why don't you get hold of him and get the details?'

Exhibit 8: Excerpt from Alcoa's Policy Guidelines for Business Conduct

---

Industrial Hygiene Protection

Maintenance of a healthy workplace in order to protect each employee from potentially harmful agents in the working environment is a basic premise underlying all Alcoa operations. This is achieved by:

• Periodic evaluations of the workplace for early detection of environmental conditions that might prove detrimental to employee health

• Institution of appropriate controls to minimize employee work exposures to potentially harmful conditions, in accordance with good industrial hygiene practice

• Training Alcoans to recognize the potential health hazards in their jobs and the means of avoiding undue exposure

• Assessing and recommending personal protection devices and providing instructions for their use and care

• Active industrial hygiene support of joint labor-management safety and health committees

• Cooperation with local, state and federal government agencies charged wth employee health protection so as to aid in the development of sound and effective standards and the development of common goals.

"The people in general engineering provide good overall design and tell the plant manager how to operate his system so he will stay below two fibers. But the small refinements in most cases have to be done locally. In many cases, these are not major expenditures or major undertakings. Our corporate engineering design people have designed to the best that they know how for a particular given time. Let's say they have designed for one-half fiber basically, with the best technology they could get. It very often happens that we get a marginal reading in a plant because of work practices."

The consultative interaction between an industrial hygiene auditor and plant management is described by Cyanamid's corporate director of safety and industrial hygiene, Vincent Caporossi, in an industry publication:

"Cyanamid has a series of detailed in-house industrial hygiene standards which describe the traditional 'detection-measurement-control' functions which form the basis for each divisional and plant program. Depending on the location, everything from chloroheptane to dipropylamine to lacquer thinner is sampled, using a combination of charcoal tubes, passive dosimeters, personal sampling pumps, and detector tubes. How often such sampling is done is detailed in the Cyanamid standards, but basically, Caporossi said, 'It depends on how much of a substance is present. If we know the amount of exposure is one-tenth of the allowable limit, we might sample twice a year; whereas if it's one-half of the permissible level, we're obviously going to do it more often.'

"When safeguards are in order, Caporossi explained, 'The decision-making comes in after the measurements are taken. Then the hygienist talks over the results with the plant management, and a collective decision is made on engineering controls, personal protective equipment, product substance change, say from a granular to a liquid form.' "[1]

Setting Up—and Improving—an Industrial Hygiene Effort

Establishing a systematic and comprehensive industrial hygiene program has been a major effort for many companies in the past few years. Finding out what materials are worked with in each operation comes first. Then, baseline monitoring for each plant location must be initiated to determine potential exposures. The company industrial hygiene function may then use an OSHA standard, or develop a company standard. It may become engaged in development of new monitoring standards and techniques.

A majority of the company environment managers visited by The Conference Board offered comments on the rapid advances in measurement technology during the 1970's. In particular, better instrumentation permitted industrial hygienists to pick out and measure smaller and smaller levels of contamination. The engineering of area monitoring equipment (and of personal monitoring or sampling devices worn by the worker) has improved its sensitivity and reliability. (See box on page 25.)

Companies have expanded their industrial hygiene efforts in several ways:

• By greatly enlarging corporate industrial hygiene staffs;

---

[1]Paul G. Engel, "Respiratory Protection: Cornerstone of this top-flight industrial hygiene program." *Occupational Hazards,* April, 1980, p. 74.

Industrial Hygiene

Industrial Hygiene is the study of workplace conditions likely to have an adverse effect on employee health and deals with potential chemical, biological and physical exposures on the job.

As manager, you should be concerned with whether or not your operations and processes:

- Present an immediate or long-term health risk for your employees and customers.
- Have the potential to aggravate a pre-existing injury or illness,
- Cause discomforts or inefficiencies that may contribute to accidents.

Good industrial hygiene calls for a "horse sense" kind of approach. The first step is to recognize a hazard. Next, that hazard must be defined as being either immediate or remote.

The third step is to measure or quantify the hazard and determine the degree of risk involved and identify those employees who may be affected. Harmful exposure to noise, toxic dusts, chemical fumes, heat, radiation and other stresses are not always easily detected. You need to determine what toxic materials or other hazards are present at significant levels even though they may not be sensed by the employee.

Fourth, hazardous situations should be corrected and controls established. This may be accomplished by substitution, containment, ventilation or elimination.

The fifth step is to establish a program of rechecks to insure your goal has been reached and that control is being maintained.

Further, you will need documentation to show ambient levels and historical variation of material concentrations of all potentially toxic materials in the work environment. This documentation also applies to physical hazards to which employees may be exposed.

Toxic Material Inventory

In order to do this, you should prepare a toxic material inventory, which is a listing of raw substances, process by-products and their location in the plant.

After establishing the toxic material inventory, you must decide how best to check on the flow of materials in the operation. For example, one way would be to review the history of a material as it passes through an entire plant.

An alternate approach is to review the operations from stationary points to see how those points are affected by materials passing through them.

The next step is to implement a proper sampling and monitoring schedule, develop exposure controls and finally document control procedures and sampling results. Because processes and technology are constantly changing, you should carefully control purchasing of toxic materials to prevent circumvention of established controls.

What all of this adds up to is an early warning system to prevent health crises from happening. An active industrial hygiene program demonstrates concern and good faith to employees, the Company, regulatory agencies and the public.

• By pressing industrial hygiene responsibility upon line management (see, for example, Exhibit 9);

• By assigning other personnel to take over operating-level responsibilities for monitoring, and for sustaining the industrial hygiene program.

Companies have also undertaken industrial hygiene training programs. These often serve to broaden in-house expertise to the operating plant level throughout a company. Taking one major product line at a time, a company can, through a major industrial hygiene program with strong technical content, systematically set standards and equip line management to be responsible for them. In addition, many companies have instituted health and environmental meetings that inform management of current standards, strategies for assuring that they are met, and the policy priority or emphasis that corporate line management needs to understand.

Corporate and divisional communications programs also rely on technical manuals. One loose-leaf "Guidelines for Work in Confined Spaces," used throughout a large energy company, provides an example. In 21 pages of clear and specific technical explanation, the corporate requirements are presented—with reasons and with examples. The guidelines follow a topical outline (Exhibit 10) and close with the ob-

servation: "The job of first-line supervision is to implement the policy contained in this document." In appendixes, the guidelines provide engineering advice for suggested ventilation practices. Other appendixes offer specifications, purchasing information, and performance characteristics for portable analyzers of confined space atmospheres, and personal monitors with alarms for continuous use in confined spaces.

The group of functions subsumed under industrial hygiene is broad and varied. One manager called it "technical housekeeping." It links together with medical surveillance to produce the strong base for all company programs in the health environment area. In Exhibit 11, industrial hygiene is threaded throughout the Allied Corporation policy statement as the foundation of company control. Just so, companies visited by The Conference Board placed heavy emphasis on the ever-present industrial hygiene programs and internal regulations (e.g., see Exhibit 9).

Exhibit 10: Considerations for Work in Confined Spaces

The following outline provides a ready reference or check list of the major considerations that apply to work in confined spaces. This generalized outline should be used in conjunction with the detailed guidelines that are presented in order to minimize the risks involved in confined space work.

I.   **Pre-Entry**

     A.   Worker Selection
         1.   Physical and psychological fitness

     B.   Worker training
         1.   Potential dangers
         2.   Safe work practices
         3.   Ventilation equipment
         4.   Personal protective equipment
         5.   Emergency rescue procedures

C. Appointment of Overall Coordinator
  1. First-line supervisor or other trained individual
  2. Responsibilities of overall coordinator
    a. Coordinates planning of work
    b. Coordinates supervising of work
    c. Implements emergency rescue plan
    d. Initiates safe work practices
    e. Posts work area
    f. Isolates confined space
    g. Evaluates confined space environment
    h. Authorizes entry by permit
    i. Provides for continued monitoring of confined space during work
    j. Suspends work or evacuates space if conditions change to present real or potential danger
    k. Assures that an outside man is available for rescue

D. Recognition of Potential Hazards
  1. Physical agents
    a. Pressure extremes
    b. Temperature extremes
    c. UV radiation
    d. Noise
  2. Chemical agents
    a. Combustible gases or vapors
    b. Toxic gases or vapors
    c. Combustible or toxic liquids or solids
  3. Oxygen deficiency
  4. Potential hazards during work

II. **Entry**

A. Isolation of Confined Space
  1. Blank flange piping
  2. Disconnect lines
  3. Lockout and tagging of mechanical and electrical equipment

B. Posting of Area
  1. Signs
  2. Barricades

C. Initial Testing of Confined Space Atmosphere
  1. Oxygen deficiency
  2. Chemical agents
  3. Physical agents

D. Comparison of Initial Test Results with Existing Standards to Determine Ventilation and/or Personal Protection Requirements

E. Ventilate and Provide Personal Protection

F. Provide Rescue Capability—including outside man suitably equipped

G. Issue Entry Permit

III. **Post-Entry**

A. Continuous or Periodic Monitoring of Confined Space Atmosphere
B. Assure Safe Work Practices Followed
C. Reissue Permit After Prolonged Absence From Area or if Conditions Change

Exhibit 11: Product Responsibility Program Policy Statement (Chemicals Company of the Allied Corporation)

Environmental/Health/Hygiene Policies

Medical Surveillance/Occupational Health

The Company will take affirmative actions to protect the health of its employees and the public from any occupational disease or other impairment which might occur as the result of our processes or products, and will provide for regular medical surveillance of Company employees.

Policy Intent

To assure:
- that all Company locations comply with all applicable industrial hygiene regulations and Company practices and that they take affirmative actions to protect our employees and the public from health-related hazards of our raw materials, products, and processes
- that periodic physical examinations of Company employees be offered in accordance with Corporate guidelines and in compliance with applicable regulatory requirements
- that previously unknown adverse health effects of products and materials are reported to the Federal Environmental Protection Agency in accordance with the Substantial Risk Regulations of the Toxic Substances Control Act
- that previously unknown adverse health affects of products and materials are disclosed to all affected employees

Product and Process Safety

The Company will make all reasonable efforts to develop and maintain information on acute hazards and environmental and toxicological effects of raw materials, intermediates, products, processes and wastes, and will take reasonable steps to eliminate these hazards or minimize the dangers posed by their existence.

Policy Intent

To assure:
- that the chemical substances handled are reviewed, existing information on hazardous properties is assembled, hazard risks are assessed, and required toxicity and environmental fate data are developed on a priority basis
- that all Company facilities receive periodic site reviews to identify and correct both chemical and non-chemical safety hazards
- that hazards are eliminated, where possible, or that affirmative actions are taken to protect employees, customers, the public and the environment
- that affirmative action programs include compliance with the Toxic Substances Control Act and all other laws and regulatory requirements

Environmental

The Company will take affirmative actions to implement and support all elements of the Corporate Environmental Policy with primary considerations for the protection of its employees, the public and the environment. These include:
- taking all practicable measures necessary to prevent or abate air, water and solid waste pollution resulting from its operations
- cooperating fully with government agencies charged with pollution control, as well as compliance with all other matters covered by Corporate Environmental Policy

Policy Intent

The purpose of this policy is to reaffirm that all practical measures will be taken to prevent or abate pollution from the Chemicals Company operations. This includes anticipation and correction of conditions or changes that could conceivably culminate in an incident that would be injurious to the employees, the public or the environment. All elements of the Corporate Environmental Policy will be implemented and supported.

Chapter 6
Toxicology

LABORATORY TESTING for adverse effects of chemicals provides one basis for hazard identification and risk assessment. The others are, generally, epidemiology and chemical analysis. Industry-supported toxicology programs occur in three settings: *multicompany* laboratories, *contract studies* with independent laboratories, and *in-house* company laboratories. A very few in-house company laboratories are decades old, but for private industry, generally, they are a development of the late 1970's. Contract toxicology has been undertaken at least since the 1940's. Development of the multicompany, association laboratory has occurred in the last decade: The Chemical Industry Institute of Toxicology was established in 1974.[1]

In toxicology testing, animals (or bacteria, or mammalian cells) are exposed to the chemical under conditions in which all other factors are rigidly controlled. The resulting changes are identified, and a "dose-response" relationship is established. The exposure may be by ingestion, by inhalation, or by contact. The testing protocol may be devised for measuring acute effects, or long-term chronic effects. (See box.)

One of the oldest in-house laboratories is Du Pont's Haskell Laboratory, opened in 1935. It is part of the Central Research and Development Department, which reports directly to Du Pont's Executive Committee.

"Haskell's responsibility is to screen compounds, conduct laboratory tests, recommend measures necessary to limit exposure to chemicals, and advise company plants, research centers, and customers of possible problems and corrective measures. It does research in toxicology and testing techniques. The staff of 208 in-

cludes 84 persons with technical degrees in fields such as animal pathology, microbiology, and medicine. A new expansion is now under way, and by the end of 1986 the total staff will number about 280.

"The laboratory performs a wide range of literature searches and animal studies to determine possible toxic effects of 900 or more chemicals each year."[2]

During the late 1970's, the pace of openings of in-house or single company laboratories increased. Monsanto opened an Environmental Health Laboratory in 1978. Mobil established a Toxicology Division that will move from rental facilities to its own site at the Mobil Technical Center in 1983. Exxon opened its own laboratory in 1980. By 1980, the Chemical Manufacturers Association identified seven chemical companies with "complete toxicological laboratories."[3] Some of the companies visited by The Conference Board indicated that their reasons for opening in-house facilities were a desire for more control over the quality of the studies and their view that analyses of specialty, proprietary chemicals are best conducted within the corporation; while tests of "commodity" nonproprietary chemicals are appropriate for external facilities to carry out. To go so far as to set up and staff a full laboratory, moreover, suggests that these companies expect to have a heavy, varied and increasing internal demand for testing over the long term. In other words, the "problem" will not go away, it will grow.

Extrapolation of test results from animal to man, from

---

[2] "Occupational Safety and Health: a Du Pont Company View," p. 16. A fuller description of Haskell Laboratory work may be found in *Haskell Laboratory,* published by Du Pont in 1980.

[1] The Chemical Manufacturers Association and the American Petroleum Institute also sponsor or contract for toxicology studies.

[3] *Chemical Product Safety: What We're Doing About It.* Chemical Manufacturers' Association, Washington, D.C., 1981.

Testing For Toxic Effects

"Toxicity has been defined as the capability of a chemical to harm a living organism. It depends on the physical and chemical properties of the compound, on the characteristics of the organism with which the chemical interacts, and, above all, on the amount of the chemical that is absorbed by the organism, that is, on its dose. The relationship between the dose and the type and magnitude of the effects and the incidence of the effects in a population are the central concerns of toxicology.

The effects also depend on the way in which the chemical is absorbed by the organism (inhalation, skin contact, ingestion, injection), how the dose is distributed in time (single dose, repeated doses, continued uptake), and on whether the magnitude of the dose is constant or variable. A deleterious effect may be caused by the parent compound or its metabolic products, which have to be identified. The transport, distribution, and elimination from the organism, both of the parent compound and of its metabolites, have to be evaluated. Effects may appear soon after exposure or may take considerable time to develop. Environmental conditions such as the presence or absence of other chemicals and the intensity of physical factors—light, temperature, humidity, radiation, and noise—may also modify the toxic action of chemicals.

All these phenomena and processes have both qualitative and quantitative aspects. Statistical correlations and mathematical models may be useful tools in toxicology but are of limited value unless their biological basis is understood, at least to some extent. To express toxicological information in quantitative terms is a complex task.

"According to some recent estimates, about 70,000 chemicals are currently used in various applications, and many of them have not yet been tested adequately, if at all. The number of chemicals is increasing rapidly; about 200 to 1,000 new ones are put on the market every year. In order to reduce the cost and time needed for toxicological assessment, an effort is being made in many countries to develop appropriate methods for screening and identifying those compounds that require long-term testing. One approach is to use rapid laboratory bioassays, such as those used in mutagenicity studies; the other is to use the relationship between chemical structure and biological activity."[1]

[1]From a book review by V. B. Vouk, of *Quantitative Toxicology*, by V. A. Filov, A.A. Golubev, E.I. Liublina, and N.A. Tolokontsev, in *Science*, October 24, 1980, p. 480.

---

one animal to another, or from one target organ to another is a source of great difficulty. The toxicological approach is challenged by the comment: "Who says it is a carcinogen? Just because it causes disease in a rat does not mean it affects humans." Since laboratory exposure of humans is unthinkable, this sort of attack on animal toxicology throws the balance of scientific inquiry out of the laboratory and back into epidemiological observations of people.

Commenting on the two methods of study, Dr. David Rall of the National Institute of Environmental Health Sciences observed:

"The classic statement is that the proper study of mankind is man. I think there is an undercurrent of feeling amongst some people that perhaps the use of animals studied appropriately or inappropriately in carcinogenicity testing is not as necessary as it is claimed to be. It seems to me that we must in fact use animal tests, today at least, as the basis for prediction of carcinogenic activity. Surveillance of the human population or selected subsets of the population for incidents of tumors is very, very important; but this is a last resort. If, in fact, an agent does enter the environment that does cause cancer in man, by the time it would be detectable in any sort of reasonable disease surveillance system, we would already have a massive epidemic of environmentally caused carcinogenesis. It is too late a point in time to have identified the carcinogen. Secondly, the manpower resources in the United States in terms of chronic epidemiology are so woefully weak in numbers—not in quality, but in numbers—that it would simply be unrealistic to view, any time in the near future, epidemiology taking on any more than it is doing right now."[4]

Since each of the disciplines of epidemiology and toxicology has a contribution to make, the integration of their results should be a prime scientific concern. Cooperative focusing on a particular set of public health concerns may, in the decade of the 1980's, become an aspect of the "professional standards" of these scientists. The professional drive to share information and goals of inquiry may override institutional particularism.

[4]David Rall "Problems of Low Doses of Carcinogens." *Journal of the Washington Academy of Sciences,* June, 1974, p. 63.

Exhibit 12: A Petrochemical Company's List of Toxicological Testing Requirements for a Single Chemical under Various Regulations and Administering Agencies

	Testing is Mandatory	Probable, Can't See How We Can Avoid	Testing Need is Possible
Food and Drug Administration			
Acute	X		
Chronic Oral	X		
Carcinogenic	X		
Reproduction	X		
Mutagenic		X	
Neurotoxicity		X	
Metabolism		X	
Environmental Protection Agency			
Acute		X	
Chronic Oral		X	
Sub-Chronic Inhalation		X	
Carcinogenic		X	
Reproduction		X	
Mutagenicity		X	
Neurotoxicity		X	
Environmental		X	
Metabolism		X	
Occupational Safety and Health Administration			
Acute	X		
Sub-Chronic Inhalation		X	
Carcinogenic			X
Reproduction		X	
Neurotoxicity			X
Metabolism		X	
Consumer Product Safety Commission			
Acute		X	

Engineering

Some companies have taken steps to include health and environmental considerations in the group of concerns or criteria addressed by engineering staffs. Moreover, companies reported that health criteria are being introduced at earlier stages of design. In this way, engineering design is not "surprised" at a late stage of development by an element that had not been taken into account.

A formal system of engineering approvals was described by the Vice President of Health, Safety and Environment of a manufacturing company this way:

"We don't do anything in this company without an appropriation. Every single appropriation request comes across my desk, so that I can review what they are going to do, and how they are going to do it. If it is under $100,000, it comes across for review. If it is over $100,000, my written approval is required. If it is a hazardous material, we brief them, alert the hygienist, inform them of protective devices they must include. If it is real bad, we just say 'no.' "

Because of the formal approval system this vice president has a more informal, early consultation with engineering departments:

"I meet regularly with the man in charge of all engineering, not only environmental engineering. We talk over the engineering solution. "Fifteen or twenty years ago," he commented: "Engineering was traditional. They were to design a production system or build buildings in the cheapest possible way that could be consistent with good output. Our engineers would deny this but it is a matter of record. They would submit a whole plan—'We're going to build such and such plant at such and such cost, and it will do this, with payback.' Then the board of directors or top management would review it, and say: 'Great plan. Now how about lopping a million dollars off it?' The engineers would lop off all the stuff that wouldn't interfere with production. This has been traditional in industry. Now they don't do this, because we don't let them do it."

Chapter 7
Staffing and Managing

IN DISCUSSING their health and environmental programs, company managers described rapid growth in staff dealing with health compliance. Comments about a tripling of staff assigned to health and environmental concerns—in one or two years—were common. A second observation is that there are many new specialists in the employ of private corporations: for example, epidemiologists, biostatisticians, toxicologists. Company managers described hiring "our first" of these specialists in the past three years.

The third development is a shift in the old specialty areas (medicine and engineering), in which public health orientation and public health skills are being imported to the occupation. The engineering field is developing specialty areas in health engineering, exposure engineering, detection-and-monitoring engineering. The field of occupational medicine is undergoing major changes in identity and viewpoint, skills required, professional orientation, and purpose.

Monitoring Staff

Growth of monitoring staffs is directly attributable to regulation. The key occupation here is industrial hygienist. The American Industrial Hygiene Association reports that about 150 large companies now employ these specialists. It estimates that 15,000 will be needed in a decade, compared with the estimated current 6,500.[1] A more conservative projection was issued by NIOSH in a consultant-prepared report in mid-1978. It estimated that about 4,800 industrial hygienists were then working, mostly in manufacturing and in government. By 1985, the report projected employment at 5,600 to 8,100. Because the supply was deemed insufficient, NIOSH set up twelve "Educational Resource Centers" for university training of industrial hygienists *and* to add health and safety elements to engineering and medical curricula.[2]

Companies adding to their monitoring staff are drawing chemical engineers into the task, training technicians themselves, adding a monitoring component to other engineering functions. One manager described the industrial hygiene staff this way: "They are basically our eyes and ears. If any of these hygienists see visible emissions coming out of a plant, they are right on it—almost by phone. They audit for everything. In the case of asbestos, federal law specifies that they are in that plant twice a year, so they *are* in that plant twice a year. Our own corporate regulations require each plant to be surveyed once a year, at a minimum." His company has 70 to 80 U.S. locations, and another 20 worldwide. By corporate policy, all are measured against the same company-established standards. In addition, these hundred locations are monitored for compliance with local standards where they exist, federal standards, and those of their host countries when these have been mandated.

A regulation-interpreter on that particular industrial hygiene staff keeps up with the standards. "He works full time reading government regulations, interpreting them, and educating our production people to be geared up for what is coming down the road. He concentrates on federal and state regulations. We have to depend on the plants for local developments as there are so many localities. Corporate keeps track of local developments in California, where corporate managers may be held criminally liable for some health derelictions. This obviously is going to have to be beefed up, simply because of government demands."

[1] *Wall Street Journal,* Labor Letter, February 17, 1981, p. 1.

[2] Paul G. Engle, "Narrowing industrial hygiene's manpower gap: Are we making progress?" *Occupational Hazards,* March, 1980, pp. 69-70.

The '007'

An unusual job, that appears to be within the "monitoring" configuration, is described by the health and safety director of a minerals company:

"When we visit an operation, we announce it two months or two weeks ahead. The senior manager knows the exact day and hour that the hygienists are going to arrive. When we get there, that plant looks b-e-a-u-t-i-f-u-l. Spick-and-span. But senior officers really want to know what their plants look like the other 360 days. So they asked us to set up an '007'—a one-man function. Not a standard hygienist monitor—he doesn't have sampling instruments. He arrives at a plant unannounced (no one knows what his schedule is), goes in, and takes photographs of anything that is irregular. It could be housekeeping; it could be the water system, dust, air emissions, safety—anything in the environmental quality area. And he makes notes.

Then, he sits down with the plant manager before he leaves the plant. He lays his pictures on the table, discusses each one with the plant manager. So the plant manager is fully aware of what he is going to get in the report. Up until that time, it was surprise and secrecy as far as when he was coming; but afte᠎ he is there, he tells them exactly what is wrong with the operation and what will be in the written report. He prepares a written report, adding the pictures with a 'before' notation. The plant manager is asked to respond within 15 to 30 days with 'after' pictures to make sure that things have been cleaned up.

Only after the second batch of pictures is taken, is the report sent to the plant manager's boss, at the divisional level. That is for the routine, the minor, things.

Occasionally '007' will uncover something that has to be changed fast. He will call me (the Vice President for Health and Environment). If I deem it serious enough, I go to the general manager or the senior officer in charge of that plant immediately and say that it requires fixing within hours. I have never run into any difficulties. They have always agreed with me, and they have always done the corrections within the proper time limits. If the investigator develops a 'feeling' that an industrial hygiene measurement should be run, he'll tell me, and we will initiate one. We all work together, keeping tabs on all of these areas, even though there are separate groups. They all feed their data to me, or in some cases they will go right next door (to the hygienists) and tell them—if it is a minor problem they will leave me out of it completely. Which is fine with me."

The corporate direction of such a monitoring operation, connected as it is to engineering, may itself have an engineering (rather than medical) tone to it. Although the interviews for this study focused solely on health, the "safety" part of the subject was found here, in industrial hygiene. (See box.) Engineering, safety *and* health are closer together in this monitoring job than in any other described. (See Exhibit 13.)

Testing Staff

Laboratory testing of substances has grown, calling for staffs of toxicologists, pathologists, veterinarians and other scientists. This increase in testing is due primarily to regulation; particularly TOSCA, but also OSHA and other health and environment legislation. So far, roughly a dozen large companies have their own full-scale toxicology laboratories in operation or in process of being established. In these companies, there is a scientific staff grouping with a configuration unique to animal testing, often organizationally isolated and in separate facilities.

In companies that "contract out" for testing, there may nevertheless be some staffing of toxicologists, biostatisticians and scientists to oversee the contract testing program. Directors of toxicological research, managing some in-house and some outside work, appear to be a fast-growing occupation within large companies.

Health Surveillance Staff

Surveillance requirements have caused companies to increase their own medical staffs, and to decrease their "distance" from the external medical-care industry. Specifically, during the 1970's, there were many first-time appointments of medical directors for companies, divisions and plant locations. The range of oversight for such positions has changed enormously.

Medical staffing has increased at corporate and at

plant levels. In-plant facilities have been opened at additional locations, staffed by an occupational health nurse and, perhaps, a part-time physician. While treating employees is the major task of plant clinics, the "growth area" has been preplacement examinations. These become the baseline for monitoring exposure and reactions to exposure. Their necessity as a defensive data base has grown with each new discovery or suspicion of occupationally related disease.

A survey done by John Short & Associates, Inc. described the existing staffing of plant clinics:

"Staffing is dominated by RN's or nurse practitioners, with a minimal number of physicians. Staffing ratio is approximately 1 physician per 2,000 employees. Nurse ratios vary by shift distribution and by the type of industry."

The consulting firm also provided these data from a survey it carried out for NIOSH:

Percentage of Manufacturing Establishments by Size with Medical Personnel

Type of Personnel	Number of Employees		
	500 Employees	500-999	1,000 +
Nurses	15%	48%	44%
Doctors	1%	2%	15%
Number of establishments in sample	1,460	248	185

Due to increased regulation and accountability, medical staffs are changing: Professional preparation and orientation is shifting toward public health. In some cases, corporate staffing may be moving away from the use of *physicians* for this new "public health" responsibility. In other cases, Conference Board interviews suggest that the traditional physician viewpoint is being expanded and changed. *What corporate physicians are trying to accomplish; what they are being held accountable for; and to whom they are being held accountable*—all of these professional job aspects are changing.

The external medical industry (physician - and hospital-based) is a service industry, oriented toward curative medicine. It provides a service—the cure of disease or disability that has already occurred. Within companies, the company-financed health unit may provide some of this service-type, curative medical product.

In the past decade, however, the differences between a corporate medical interest and the medical service industry's interests have become acute. Occupational physicians seem to be passing through some new professional territory, and to be struggling with the subtleties of professional identification. Some aspects of the struggle are not new (see box), but are far more difficult because government regulation has intervened.

In Henrik Ibsen's *An Enemy of the People*[1] (1882), Dr. Stockmann, medical director of a resort town's medicinal baths, finds out that they are polluted and the cause of typhoid among visitors. His brother, Peter Stockmann, is the mayor and chairman of the baths.

In Act II:

"Mayor Peter Stockmann. As you have been so indiscreet as to speak of this delicate matter to outsiders, despite the fact that you ought to have treated it as entirely official and confidential, it is obviously impossible to hush it up now. All sorts of rumours will get about directly, and everybody who has a grudge against us will take care to embellish these rumours. So it will be necessary for you to refute them publicly.... After making further investigations, you will come to the conclusion that the matter is not by any means as dangerous or as critical as you imagined in the first instance.... And, what is more, we shall expect you to make public profession of your confidence in the Committee and in their readiness to consider fully and conscientiously what steps may be necessary to remedy any possible defects.

Dr. Stockman. But you will never be able to do that by patching and tinkering at it—never! Take my word for it, Peter; I mean what I say, as deliberately and emphatically as possible.

Mayor Peter Stockmann. . . . as a subordinate member of the staff of the Baths, you have no right to express any opinion which runs contrary to that of your superiors.

Dr. Stockmann. This is too much! I, a doctor, a man of science, have no right to —!

Mayor Peter Stockmann. The matter in hand is not simply a scientific one. It is a complicated matter, and has its economic as well as its technical side."

[1]Act II, trans. R. Farquharson Sharp, New York: Everyman's Library, Elsevier-Dutton Publishing Co., Inc., 1967.

There has been, for example, a good deal of concern with the duty to inform—to inform fellow employees of the company, to inform (or alert) the governmental authorities, to inform other professionals outside the company's employ. The American Occupational Medical Association, in its first "code of ethics" in 1976, made some general observations in items 8, 9 and 11 (see Appendix C). Development of these phrases into specific guidance about the duty to inform, however, has been superseded by regulation. OSHA has filled the silence with orders, particularly the requirement that companies make medical records available to employees; and NIOSH, through recent court decisions, has "ordered" corporate physicians to inform outside epidemiologists (see page 18).

The professional transformation of occupational medicine into occupational and environmental public-health practice is (to a great extent) a matter of professional training. Exhibit 14 indicates the "new" requirements for corporate medicine. This doctor is unlike the private practitioner of today—this doctor is a particular type of medical detective. His or her counterparts will not be found in private practice. However, counterparts do exist at institutions like the Center for Disease Control in Atlanta, other sections of the National Institutes of Health, and medical schools with public health divisions.

The American Occupational Medical Association has begun training conferences focusing on epidemiology, to develop some professional acquaintance and skill in this area for currently practicing corporate physicians. (See Exhibits 15 and 16.)

Management Structure

The subject of environmental health receives strong corporate guidance in the large, environmentally sensitized companies visited by The Conference Board. These companies are being held responsible for producing a product—conservation of public health—that is not sold on the market. The responsibility is imposed at the corporate level by legislation, by court opinions fixing liability, and by public opinion. Since the "exposure" is corporate, the response is managed from a corporate level.

In a thoughtful essay on how the organizational structure of large corporations impedes responsiveness to social legislation (such as OSHA, TOSCA, RCRA) Robert Ackerman, vice president, Preco Corporation observes: "Chief Executives are decidedly reluctant to usurp the responsibilities assigned to division managers, knowing that such interventions will reduce the division managers' accountability for results." Since financial results for a division—for a plant—are the criteria of success, there is no incentive for managers to address the environmental externalities of the operation. Dr.

Exhibit 14: An Advertisement Specifying Requirements for a Company Medical Director

> ### Company Medical Director
>
> Excellent opportunity for health professional to direct the medical programs for two of the corporation's existing companies. Responsibilities will also include an Epidemiology Review Program.
>
> You must be licensed or eligible for licensure to practice medicine in the state and have a minimum of five years current experience in an occupational or related medical assignment. Experience with clinical aspects of lead toxicity desirable. Certification in Occupational Medicine and/or the American Board of Internal Medicine, or the American Board of Preventive Medicine required. Training in Public Health—Epidemiology required. Previous hospital medical staff membership highly desirable. Knowledge of OSHA and Workmen's Compensation required. Travel required.
>
> We offer an attractive salary, benefits package and a liberal relocation allowance. Plus the opportunity to contribute in a meaningful way to the ongoing development of the Company's comprehensive employee health system.

Ackerman notes that corporations *have* learned to respond to "social demands," and points out three elements of this accommodation:

(1) Top corporate officials take an active (and "unnatural") role.

(2) Technical help is provided, to teach the organization how to manage its response.

(3) Operating management's performance evaluation system incorporates performance on the "social issue."[3]

Corporate-level responsibility is located within the corporate relations array, frequently alongside employee relations and also various "compliance" management systems. Exhibits 17, 18 and 19 depict variants of this formation. A second location for the environmental and health subject is *within* the employee relations function, as in Exhibits 20 and 21. The element that distinguishes

[3]Robert W. Ackerman, "The Organizational Environment and Ethical Conduct in Occupational Medicine: a Perspective." *Bulletin of the New York Academy of Medicine,* September, 1978. (Full paper in Appendix F.)

Exhibit 15: Excerpts from the Annual Conference Program of the American Occupational Health Association

---

Advance Program for Physicians
1981
AMERICAN OCCUPATIONAL HEALTH CONFERENCE

Epidemiology

Epidemiologic techniques have often identified the causes of occupational disease, defined the risks of exposure and provided the basis for developing regulations and public health practices. The complexity of the modern workplace, the mobility of workers, and the absence of records have all placed limitations on the information that can be obtained by epidemiologic study. Recognition of these strengths, the weaknesses, and the possible solutions to problems in occupational medicine will be the topic of this afternoon's seminar.

Lectures:

EPIDEMIOLOGY IN THE WORKPLACE—ITS STRENGTH AND LIMITATIONS

PROBLEMS IN CHARACTERIZING THE POPULATION AT RISK

EPIDEMIOLOGIC PROBLEMS ASSOCIATED WITH EXPOSURE TO SEVERAL AGENTS

MISSING RECORDS IN OCCUPATIONAL EPIDEMIOLOGY

EPIDEMIOLOGY

Epidemiology is the study of the distribution and determinants of disease in human population. Adverse health effects of occupational hazards generally are assessed by epidemiologic methods. Although some methods may appear complex, the basic ideas usually are not highly theoretical or mathematical. This course is designed to provide the occupational physician with knowledge about basic epidemiologic methods used in the design and analysis of studies of occupational hazards, including the limitations as well as the strengths of those methods.

The course will include consideration of the following: (a) posing the question, i.e., how does one decide when an epidemiologic study is warranted; (b) what study designs are available and what kind of answers can be expected from these designs; (c) what analytic methods are most often used; and (d) what does all that statistical jargon really mean in terms of the questions facing the occupational physician. Illustrative material will focus on current topics using ongoing as well as published studies.

IMPACTING AN ORGANIZATION'S MANAGEMENT

Occupational health professionals must be able to work effectively with managers in organizations which they support. To bridge the gap between the professional's work and the manager's world it is often necessary to put health issues in perspective in relation to other management concerns. Doing so effectively requires mutual understanding. To have organizational impact, the occupational health professional must be able to work with management and may have to take the first step. This seminar will focus on the four key aspects of the health professional - line manager interface shown below. Participants will develop individual skills in a "working meeting" atmosphere. For this reason, this group will be held to a relatively small size. Topics to be covered include:

1. Management concepts, language and decision-making processes.
2. Development of organizational influence as an individual and as a staff department leader.
3. Upward communication: the creation of effective proposals, presentations, and reports.
4. Instituting organizational change.

---

these two organizational placements seems to be the relative emphasis the corporation places on its external environmental impact (compared to internal, work force impact). "Outside" environmental impacts are addressed by such laws as the Toxic Substances Control Act; the Resource Conservation and Recovery Act (particularly the regulations covering waste disposal); the Clean Air Act. Product liability considerations, also, are part of the *external impact* of the company's operations. Where these external-to-the-work-force effects are given great weight, they may pull the locus of the corporate environmental and health control outside employee relations.

Locating the function within employee relations

Exhibit 16: Example of In-Service Training for Company-Employed Physicians

THE SECOND ANNUAL

EPIDEMIOLOGY FOR THE OCCUPATIONAL PHYSICIAN

February 23 – 27, 1981

The Broadwater Beach Resort Complex, Biloxi, Mississippi

Presented by
AMERICAN OCCUPATIONAL MEDICAL ASSOCIATION

and the
NATIONAL HEART, LUNG AND BLOOD INSTITUTE

A 4 day seminar in pleasant surroundings offering practical epidemiology concepts and applications.

This postgraduate seminar is specifically designed for the occupational physician. A faculty/student ratio of 1 to 10 has been established to facilitate individual understanding. Each major presentation is followed by a small group session to provide content mastery. Applicable information provided will include: how to design and carry out an epidemiological study; measures of disease risk that can be applied to groups; how screening and detection of occupational disease can be accomplished; methods of evaluating epidemiological studies; using epidemiological findings in occupational standards; requirements for an occupational health surveillance system and how such data may be used.

PROGRAM FOR MONDAY, FEBRUARY 23, 1981

3:00-6:00	Registration
6:30	Gulf Shore Cocktails
7:30	Get Acquainted Banquet
8:30	Seminar Orientation

PROGRAM FOR TUESDAY, FEBRUARY 24, 1981

8:00-8:45 . . . Lecture . . Introduction to Basic Concepts, Measurement and Uses of Epidemiology

8:45-9:00 . . . . Discussion/Coffee
9:00-11:45 . . . . Small Group Sessions
      "Back to Basics"
11:45-1:00 . . . . Lunch
1:00-1:45 . . . Lecture . . Basic Statistics for Epidemiological Studies
1:45-2:00 . . . . Discussion
2:00-4:45 . . . . Small Group Sessions
      "What To Do With Tables--Graphs and Regressions"

PROGRAM FOR WEDNESDAY, FEBRUARY 25, 1981

8:00-8:45 . . . Lecture . . Cardiovascular Disease Risk Factors and Interventions
8:45-9:00 . . . . Discussion/Coffee
9:00-11:45 . . . . Small Group Sessions
      "Surveys, Trials, and Experimental Designs-- What Works Where"
11:45 . . . . FREE TIME

PROGRAM FOR THURSDAY, FEBRUARY 26, 1981

8:00-8:45 . . . Lecture . . Evaluation of Human Cancer Risks
8:45-9:00 . . . . Discussion/Coffee
9:00-11:45 . . . . Small Group Sessions
      "Zeroing in on Basic Methods Used in the Design and Analysis of Epidemiologic Studies"
11:45-1:00 . . . . Lunch
1:00-1:45 . . . Lecture . . Health Surveillance System for Cancer Risks
1:45-2:00 . . . . Discussion
2:00-4:45 . . . . Small Group Sessions
      "Data--What You Need, What You Can and Can Not Do With It"

PROGRAM FOR FRIDAY, FEBRUARY 27, 1981

8:00-8:45 . . . Lecture . . Uses of Epidemiological Findings in Occupational Standards
8:45-9:00 . . . . Discussion/Coffee Break
9:00-11:30 . . . . Small Group Sessions
      "Checking out Evaluation, Validity, Causality, Hypothesis Testing"
11:30-12:30 . . . . Seminar Summary, Future Plans, Student Evaluation of Seminar

suggests a heavy—if not exclusive—emphasis on employee health environment. There may be a separate unit for external environment concerns (see Exhibit 21.)

Another reason for locating the function within employee relations may be the sequence of regulatory development. The pressure of OSHA (and of unions and adverse publicity) focused public concern on *employee* exposure by the early 1970's. Attention to the environmental effects of industrial activity on air, water, soil and product users has been building more slowly over the past decade. A corporate concern might then have evolved and broadened from (1) employee safety, to (2) employee health, to (3) product usage and effects, to (4) total environmental effect of the corporation's activities.

Exhibit 17: Johns Manville's Health and Environment Organization

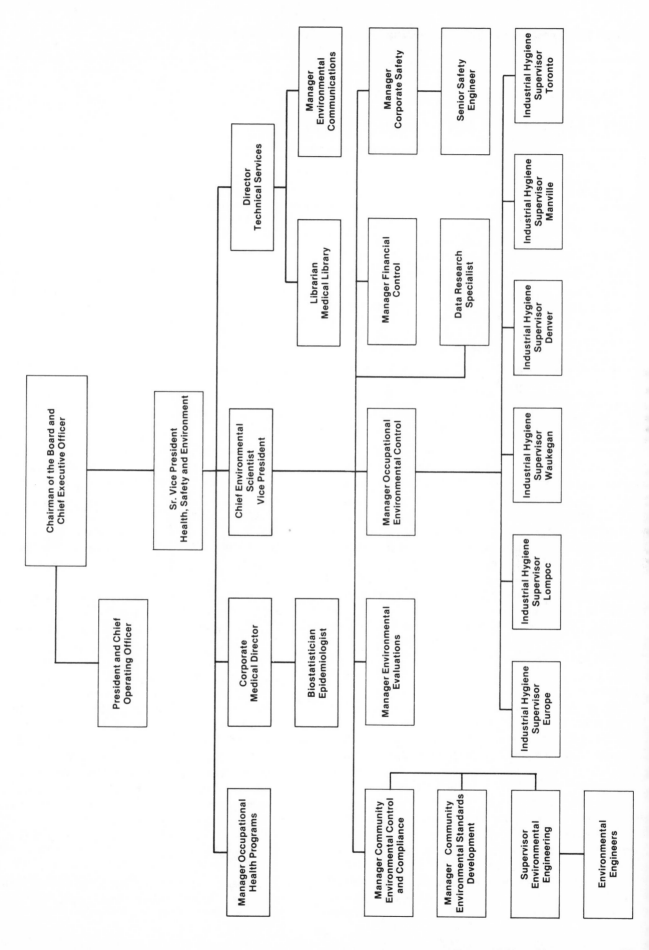

Exhibit 18: Exxon Corporation's Health and Environment Organization

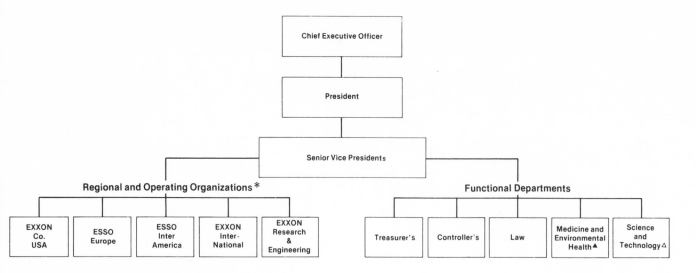

Notes: ✳ Each regional and operating organization contains subsegments responsible for discharging that organization's health and environmental responsibilities.
▲ (Includes the Research and Environmental Health Division)
△ (Includes an Environmental Affairs Coordinator)

When this broadening of obligations occurs, the company might respond by opening new organizational entities dealing with health effects in product development, in plant siting and engineering, in transportation, in marketing, and so on. However, this does not seem to be the way the system is developing in most of the companies visited. Instead of such fragmentation, an aggregation of health impacts is being addressed by unitary corporate-level attention. The impetus for such unification may be that poor performance exposes the whole of the company, with pervasive effects. For example, an insecticide can be an employee health, waste disposal, environmental, transportation, and product-user problem. Yet dealing with it separately in each appropriate corporate activity may not be effective or reasonable.

The kinds of technical attention that various environmental and health hazards require are very similar. For example, toxicology testing is likely to serve needs that could be semantically isolated—as "occupational health," "product development," "environmental effect." But both effectiveness and efficiency, and the scarcity of technical resources, will lead to one company toxicology laboratory, not three. Similarly, the industrial hygiene inspections of a company visited by The Conference Board cover worker exposure, stack emissions, waste treatment, and so on.

A few companies reflect this "technical integration" of the subject in their organization; others accomplish

integration through a committee mechanism. For example, in Atlantic Richfield the health and environment subject is within a technology configuration (Exhibit 22.) The Allied Corporation has also fully integrated the oversight of internal and external impact (Exhibit 23,) and made the function answerable to an Environmental Policy Committee of the Board of Directors. This committee is comprised of outside directors.

Committee and task force organizations are used, in some companies, to integrate health and environment obligations with other operating concerns. "Integrate" is preferred to "impose." Nevertheless, since the obligation is often a legal imposition—a regulation—these organizations have oversight roles as well. For example, Atlantic Richfield uses middle-management panels, directed by a corporate-wide Health, Safety and Environmental Protection Council, to review environmental protection systems in major operating facilities. The program is generally described in Exhibit 24. A separate program, similarly organized, reviews occupational safety and health performance and systems, for the Council. The melding of internal and external health and environmental concerns is at the corporate level, with the particularly "technical" flavor depicted in Exhibit 22. The company explains:

"Because of the remarkable overlap of environmental, safety and health issues, our corporation in 1979

Exhibit 19: Du Pont Corporation's Health and Safety Organization

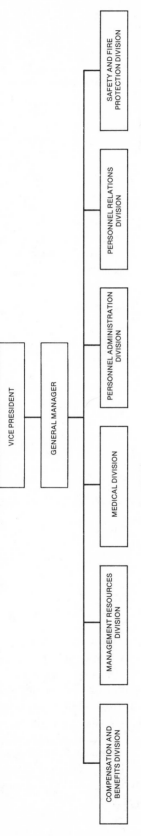

VICE PRESIDENT

GENERAL MANAGER

- COMPENSATION AND BENEFITS DIVISION
- MANAGEMENT RESOURCES DIVISION
- MEDICAL DIVISION
- PERSONNEL ADMINISTRATION DIVISION
- PERSONNEL RELATIONS DIVISION
- SAFETY AND FIRE PROTECTION DIVISION

ADMINISTRATIVE

Formulates Corporate medical policy, and audits for compliance, provides consultation, renders medical service or advice, and acts in an advisory capacity to all medical units throughout the Company on a world-wide basis.

Represents the Company on several organizations such as the AOMA, AAOM, etc. Assists with presentation of Company position to Federal and State safety and health organizations.

OCCUPATIONAL MEDICINE CONSULTATION

Provides consultation on occupational medicine to all Company units. Audits plant occupational medical programs for compliance with Corporate policy and guidelines.

EPIDEMIOLOGY

Advises on all matters pertaining to medical recording and the application of statistical methods to the study of medical records and employee health. Collects data necessary for epidemiology studies and conducts these studies.

Coordinates studies being done on contract. Monitors studies being done by others and evaluates the results.

TECHNICIANS

Performs certain functions (taking blood, blood counts, urinalyses, X-rays, electrocardiograms, pulmonary function, audiograms, eye tests, etc.) with respect to pre-employment, periodic and special physical examinations.

PHYSICIANS

Conducts pre-employment, periodic and special physical examinations. Treats emergency cases and advises employees on personal health matters. Advises management on health problems. Serves as medical representatives on special committees.

ADVISOR ON ALCOHOLISM

Administers Company alcoholism program under supervision. Serves as consultant to Company locations on alcohol problems. Represents the Company on several organizations concerned with alcoholism.

NURSING STAFF

Performs first aid and routine nursing duties under appropriate medical supervision.

Coordinates and advises on risk management policies and practices with the broad objective of elimination and/or reduction of Company risks and exposures.

Advises on safety and fire protection engineering with regard to new construction and alterations of existing facilities, consults on safety and fire protection, industrial hygiene and occupational health problems, and consults on safe distribution of hazardous materials. Investigates serious accidents, fires and explosions resulting in personal injuries, loss of life, or property damage. Initiates action in the development of engineering standards for safety and fire protection. Interprets and advises on application of Federal safety, health and transportation and distribution standards.

Conducts safety and fire protection and safe distribution of hazardous materials surveys and occupational health surveys of Company and Subsidiary units. Recommends improvements in safety and fire protection facilities and programs and makes necessary follow-up to effect compliance. Provides underwriting service on fire protection engineering for all insurable assets of Company. Advises with respect to protection, disaster control and security on Company plants. Advises about regulatory compliance for off-plant shipment of hazardous materials. Serves as liaison between the Company and outside agencies on matters of protection, security and Civil Defense. Consults and audits compliance with Corporate Personnel Environment Record System.

Promotes interest in the Company's safety and fire protection program. Publishes "Managing Safety" newsletter and other Division publications.

Administers the Reserve for Insurance—Occupational Disability, coordinates and advises on workers' compensation, prepares corporate self-insurer's status reports for state workers' compensation boards, initiates and follows to completion compensation payments for permanent disability and fatality cases.

Maintains liaison with Insurance Section, Finance Department for exchange of information necessary to promote Corporate risk management objectives. Accumulates loss data, analyzes and prepares statistical reports for management use and for outside agencies such as NSC and CMA. Maintains records and publishes motor vehicle accident statistics to assist departments in promotion of safe driving.

Maintains records on all on-the-job injuries and illnesses and all off-the-job lost time injuries of Company personnel. Prepares statistical reports on safety performance and confirms eligibility of attainments in "Board of Directors Safety Award Plan." Acts as final classification authority for all injuries to Company personnel.

Members represent the Company on several committees of national organizations, such as NFPA, CMA, NSC, etc., and coordinate responses to Federal safety and health organizations.

Exhibit 20: Northrop Corporation's Safety and Health Organization

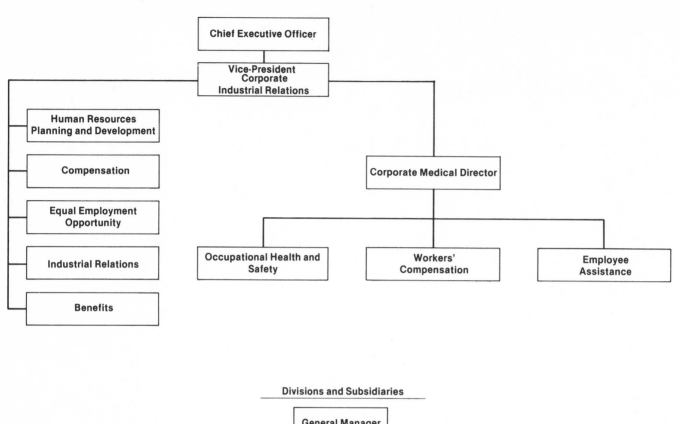

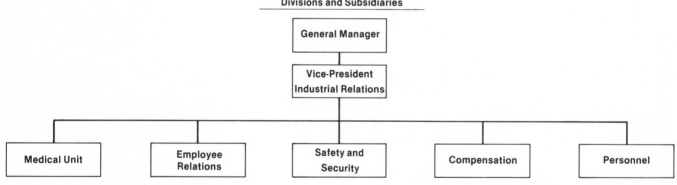

reorganized most of these functions into a single Health, Safety and Environmental Protection Group within Technology. Not only are these groups concerned with responsibilities for the traditional dimensions of environmental, safety, medical and industrial hygiene efforts, but we also established a new headquarters function of Toxicology and Product Safety."

Allied Corporation has an unusual committee system "to effectively reduce or eliminate" an employee's "legal exposure." Both the Toxic Substances Control Act (TOSCA) and the Consumer Product Safety Act (CPSA) oblige employees to report to the government if they have any information indicating that materials that are manufactured, processed or distributed by their company may represent a substantial risk to health or the environment. Allied has set up a committee, inside the company, to which the employee may report first. The committee evaluates the information and decides whether or not to pass it on to the government—but the employee is no longer liable for not telling the government. Allied describes the committee as including "representatives from within Corporate Environmental Affairs, including Medical Services, Occupational Health, Pollution Control and Product Safety, as well as from the departments of medical affairs and law."[4]

[4] *Allied Chemical today,* October, 1978, p. 2.

Exhibit 21: General Electric Company's Health and Environment Organization

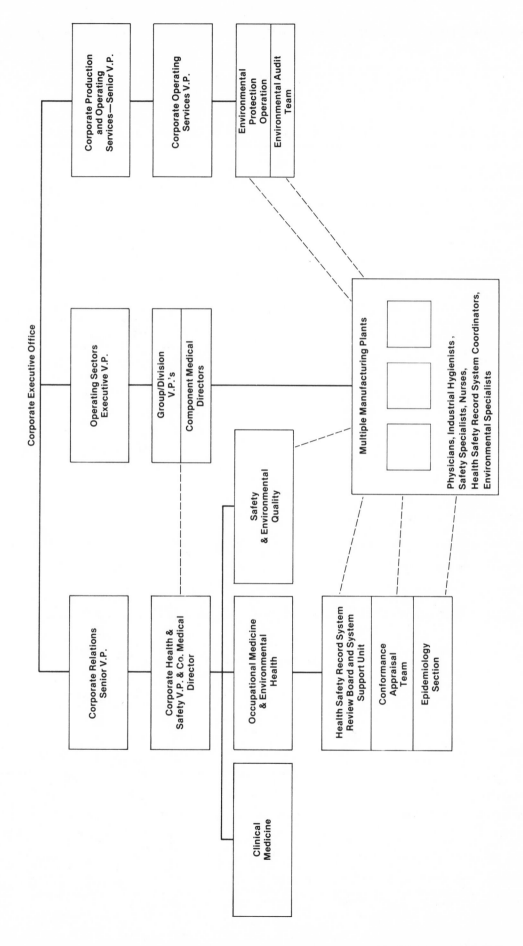

Exhibit 22: Atlantic Richfield Company's Health, Safety and Environmental Protection Organization

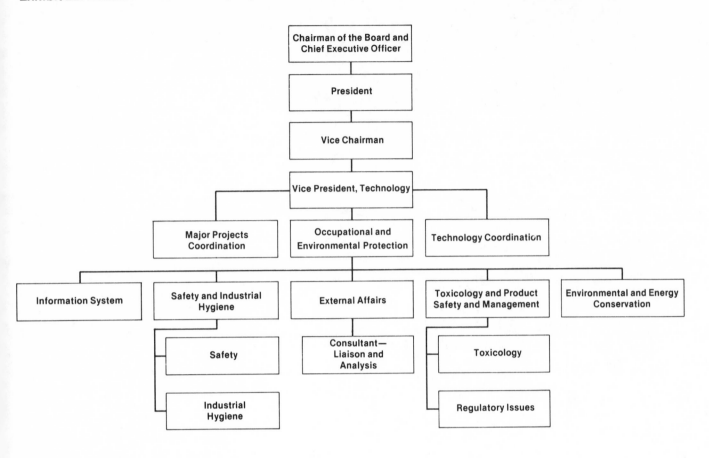

Exhibit 23: Allied Corporation's Environment and Health Organization

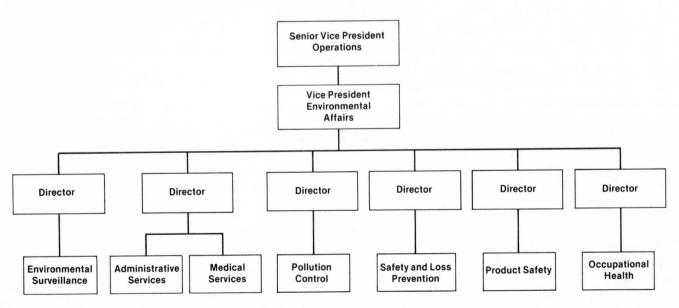

Exhibit 24: Atlantic Richfield Company's Environmental Protection Systems Review Program

INTRODUCTION

This manual describes the environmental protection systems review program and explains how it is conducted. The purpose of the environmental protection systems review program is to assess and improve overall environmental protection performance of the Atlantic Richfield Company.

It is Atlantic Richfield Company policy that each line manager is responsible for the environmental performance of his or her activity, and that achievement in protecting the environment is regularly measured. The review program supports this policy by defining what managers and supervisors are accountable for and by providing a mechanism for measuring achievement.

OBJECTIVE

The primary objective of the review program is to help managers employ management systems which achieve fullest possible compliance with the Atlantic Richfield Company environmental protection policy through periodic appraisal of major operating facilities.

SCOPE

The review program's particular focus is the system of defenses needed by a facility to protect the environment. The program's concern is the adequacy of systems for environmental protection in facility design, operating procedure, maintenance and emergency preparation, demonstrated by compliance with the Atlantic Richfield Company environmental protection policy.

The review program is built around ten criteria for evaluating management systems for environmental protection. These criteria define what is expected of managers in ten areas that are critical and provide the basis for recommendations for improvement.

PROCEDURE

The program is implemented by teams drawn from an intercompany panel comprised of more than fifty middle management employees. Panel members are nominated by the heads of the operating companies. Formal appointment is made by the corporate Health, Safety and Environmental Protection Council. The panel membership is selected to provide a mix of engineering, operating and environmental backgrounds to insure a balanced peer review of the facility.

Small teams are commissioned from the intercompany panel membership by the corporate Council for specific review projects. The review process proceeds in the following manner:

• The Council selects a facility for review and appoints a leader and team members from the panel. Teams typically have three or four members including a member from the corporate environmental staff.

• With assistance from the corporate area environmental director, the team leader arranges with the facility manager and other team members for the time of appraisal and for preparatory work needed before starting. The team leader visits the facility, requests any needed company and corporate staff support and confirms review agenda with the facility manager.

• The team assembles at the facility and conducts the appraisal which is aimed at systems designed to achieve environmental protection results identified in the ten criteria. The team works closely with local management, striving for a cooperative approach to identify and evaluate deficiencies.

• Following the appraisal, and before departing, the team discusses its observations, findings and recommendations with local management. The team presents a preliminary written report to the manager responsible for the development of the action program and response to the report.

• Approximately forty-five days later, the facility manager presents a conclusive response to the team leader. This report may state exceptions the manager may take to specific findings, gives the status of corrective actions that have been taken, and spells out the schedule for additional corrective actions which will be taken.

• The team leader reviews the facility response, edits the team report where desirable and comments on any items in contention. The facility response is affixed to the team report with comments (if any) and forwarded to the next higher level operating company manager.

• The next higher level manager reviews the report and forwards it through appropriate channels including the operating company environmental protection group, to the operating company president for review and subsequent release to the chairman of the Council.

The overall program is summarized in the following diagram.

**Environmental Protection Systems
Review Program**

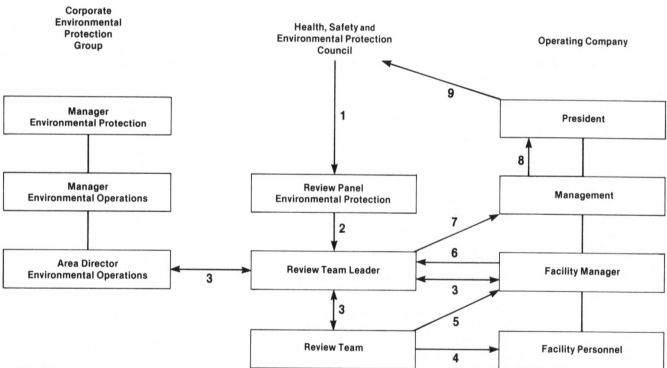

1. Select site
2. Appoint team
3. Pre-review
4. On-site review
5. Findings—recommendations
6. Facility response
7. 8. 9. Final report

Part III

Union Involvement in Occupational Health

Chapter 8
The Extent of Union Activity

THE UNION AGENDA in the United States has been wages and working conditions. The concept of "working conditions" has broadened considerably since the campaign for the eight-hour day at the turn of the century. In industries with high accident rates, such as mining and railroading, union representation began to include the introduction of safety rules and special union-management safety committees. The emergence of occupational health as a distinct subject for union representation developed at about the same time as the Occupational Safety and Health Act of 1970.

Some individual unions and the AFL-CIO supported the legislation, and union involvement increased during the regulation-writing surge of the late 1970's. Professor Thomas Kochan of M.I.T. describes the unions' aim this way:

"They supported the passage of OSHA in the belief that it would (1) provide adequate standards and enforcement, (2) serve as a source of increased power in negotiations, and (3) provide a stronger base for influencing management through joint committees. Occupational safety and health is therefore an excellent example of an issue in which public policy plays an important role as a source of power in collective bargaining."[1]

Unions, like others, first focused on safety. Occupational health began to develop into an issue only in the late 1970's, and then only in those unions with large membership groups in the chemicals and metals industries. Because of this heavy—if not exclusive—attention to safety, the number of joint committees established under major contracts offers spurious evidence of union involvement. In short, there is little information on what these safety-and-health committees actually do.

One 1976 study of local union-management safety and health committees uncovered 244 "major changes made in the plant"—mostly safety items. Asked to name the "influence" that caused the major change, management cited OSHA most often, and "management initiative" next most often. But, for those changes clearly categorized as health oriented, union influence was cited nine times; OSHA and management initiative only four times each.[2]

A 1976 study of union resources devoted to health programs concluded that the subject was "not a top priority issue." The study covered 15 unions that represent over seven million workers "particularly prone to chemical health hazards because of the industries they work in." The study found few health experts on union staffs: seven industrial hygienists, one epidemiologist, four public health specialists, one full-time doctor, four part-time doctors, two chemists, two engineers—among the 15 major unions. The Mine Workers and the Oil, Chemical and Atomic Workers had the highest health staff ratio: one per thirty thousand members. The researchers reported that union plans to expand staff and financial resources in health expertise "seem minimal." The study concluded that unions would not be able to take the initiative until they increased their own scientific research ability, and devote "a sizable chunk of (their assets) toward health research and advocacy."[3]

[1] Thomas A. Kochan, *Collective Bargaining and Industrial Relations*, Homewood, Illinois: Richard D. Irwin, Inc., 1980, p. 360.

[2] Thomas A. Kochan, Lee Dyer, and David B. Lipsky, *The Effectiveness of Union-Management Safety and Health Committees*. Kalamazoo, Michigan: The W.E. Upjohn Institute for Employment Research, 1977, Table 3-8, pp. 69-70.

[3] The Bureau of National Affairs, Inc., *Facts for Bargaining*, Part 2 of *What's New in Collective Bargaining Negotiations and Contracts."* #818, October 7, 1976.

Employees themselves may not have great confidence in union representation on health protection. A Chamber/Gallup survey in 1979 asked: "If you were a worker in a manufacturing plant, of the groups noted on this card, which would you trust the most to decide a difficult issue regarding workers' safety and health?" Union officials were named by only 6 percent of respondents. Federal health and safety agencies were named by 7 percent. Management was named by 9 percent, "fellow workers" by 21 percent. The greatest trust—40 percent of responses—was placed in a team of workers and management.[4]

Union programs have concentrated on developing union staff awareness of occupational health problems in specific industries, and in spreading that awareness to the membership. For example, in 1980 the Amalgamated Clothing and Textile Workers Union held an occupational health conference for 120 delegates, jointly sponsored by the Mt. Sinai School of Medicine. The union runs its own health care program in a number of cities; the doctors in charge of these Health Centers came to the meeting to learn more about occupational diseases.[5] Another example is a series of educational conferences run by the Paperworkers "to prepare local members to assume the tasks of Safety and Health Committees, to learn to deal with hazardous work situations, and to develop local unionists' expertise about their workplaces." The union expects these programs to "help stir local activism."[6]

Some of these union programs have been funded by OSHA grants. OSHA began a "New Directions" program in 1979 that granted at least $3.5 million to 66 "business, employee, and educational organizations" to undertake training and educational projects, and to serve as "resource centers." The idea is linked to concepts of responsibility and self-help: If individual employees (including supervisors and managers) are better informed about the hazards that may be found in occupational environments, those individuals are more likely to actively protect themselves and others.

Such programs grew during the late 1970's during the Carter Administration. During the early and mid-1970's, OSHA had concentrated its efforts elsewhere (especially on the institution and administration of detailed safety regulations) and relied less on educating employees and management. During 1975, for example, it gave George

Washington University $578,000 to make a 30-minute television show on occupational cancer for showing in 10 cities; the University of Wisconsin received $542,860 to train employees "primarily from 12 unions representing high hazard industries." The grantees of the mid-1970's were likely to be universities, states and professional associations.[7]

Bargaining over occupational health protection grew during the decade. Yet the negotiation results, as reported, are heavily procedural in tone. For example, in 1973 the Auto Workers and General Motors negotiated a memorandum of understanding that the company would pay "health and safety representatives" in plants with over 600 employees. The union names these individuals; then GM trains them (in hazard-recognition techniques and OSHA standards), pays their wages, and supplies the necessary testing equipment.[8] Inspections are conducted by company representatives and the "health and safety representative." When the representatives are in agreement that a condition is dangerous, they can shut down an operation.

In 1980, the Institute of Industrial Relations at the University of California prepared a guidebook on health and safety bargaining for union representatives. The guidebook advises:

"Historically most employers have resisted negotiating health and safety language with union respresentatives. Their position has been that the health and safety of employees is solely and exclusively a 'management prerogative'; and that management, by virtue of workplace ownership, is responsible for everything that goes on at the worksite."[9]

However, the same author says, a National Labor Relations Board decision in 1966 made health and safety a mandatory subject of bargaining; and OSHA gave it a special boost by mandating health and safety standards in all work places.

The coverage of suggested topics in this guidebook, shown in Exhibit 25, outlines a prominent role for the union in employee health protection. Many of the subjects suggested are also addressed in OSHA regulations—recommended contract language might interact with the regulations to expand and particularize a

[4]Survey Research Center, U.S. Chamber of Commerce, *Worker's Attitudes Toward Productivity,* 1980, publication 6282, Figure 2.

[5]*ACTWU Labor Unity,* Amalgamated Clothing and Textile Workers Union, New York: July, 1980, pp. 3-4.

[6]*The Paperworker,* United Paperworkers International Union, Flushing, New York, December, 1980, p. 8.

[7]*The President's Report on Occupational Safety and Health,* Washington, D.C.: Government Printing Office, 1975, pp. 73-84.

[8]Lawrence S. Bacow, *Bargaining for Job Safety and Health.* Cambridge, Mass.: M.I.T. Press, 1980, pp. 61-62. A concise and thoughtful description of the representatives' functions is provided on pages 62-67.

[9]Paul Chown, *Workplace Health and Safety: A Guide to Collective Bargaining.* Berkeley: University of California Press, 1980, p. 1.

Exhibit 25: Suggested Topics for Union Bargaining

1. General duty clauses (company's overall obligation)
2. General duty to bargain—union recognition clauses used for health and safety
3. Sanitation, housekeeping and specific working condition
4. Lighting, ventilation and noise
5. Protective clothing and equipment
6. Crew size, working in isolation, and excessive weight lifting
7. Protection from hazardous, dangerous work or from unsafe materials or processes
8. First aid, shop medical care, and other arrangements for medical treatment at the worksite
9. Regular rest periods and relief operations
10. Medical surveillance and checkups
11. Workplace monitoring for potentially dangerous health conditions
12. Payment for time lost from work while undergoing medical examination or treatment
13. Payments to supplement wage losses caused by injury or illness on the job
14. Rate retention, job transfer, and seniority rights
15. Benefit protection for workers with long-term job injury or illness
16. Worker training in health and safety
17. Workers right to refuse temporary transfer to jobs
18. Reporting of accidents or illness
19. The right of a worker to refuse to work if a job is unsafe or hazardous
20. Pay and job right issues involved when a job is shut down over safety and health problems
21. Prohibiting retaliation against any worker who engages in health and safety activities
22. Prohibition against speed up
23. A worker's right to lock out or "tag" a machine or work process that is dangerous
24. Special "vacation" or time off from work with pay when exposed to special hazards
25. Hazard pay for performing dangerous or unhealthy work
26. Information to the union
27. Union access to the workplace
28. The union's right to bring in outside consultants on health and safety
29. Employer furnished monitoring equipment
30. Walkaround pay and compensation payment for time spent handling health and safety matters
31. An employer-funded health consultant or independent research study
32. Payment for training of union health and safety representatives
33. Union's right to shut down any unsafe or hazardous work process
34. Provisions for recognizing organized health and safety structures
35. Prompt settlements of disputes over health and safety
36. Non-liability for health and safety problems

general OSHA rule. In one subject area, refusal to work under unsafe conditions, the OSHA regulation creates a "worker right" previously provided only to unionized employees.

Refusal to Do Hazardous Work

An OSHA regulation protects an employee who refuses to do dangerous work from employer-imposed sanctions. The language is restrictive:

"If the employee, with no reasonable alternative, refuses in good faith to expose himself to the dangerous condition, he would be protected against subsequent discrimination. The condition causing the employee's apprehension of death or injury must be of such a nature that a reasonable person, under the circumstances then confronting the employee, would conclude that there is a real danger of death or serious injury and that there is insufficient time, due to the urgency of the situation, to eliminate the danger through resort to regular statutory enforcement channels. In addition, in such circumstances, the employee, where possible, must also have sought from his employer, and been unable to obtain, a correction of the dangerous condition."[10]

The strategically important point in this regulation is that the right is an *individual* right. Under the National

[10]29 Code of Federal Regulations, 1977, 12 (b) (2).

Labor Relations Act, *concerted refusals* to work are protected. Group action aimed at correcting health hazards has been included by a number of court decisions, and is considerably less restricted than the OSHA rule quoted above.[11] Yet the refusal (or the protest) must meet several tests of being a group activity—on behalf of a group.

The employee acting alone is only narrowly protected. However, in extreme situations this action can arouse a union to adopt an occupational health priority that it may not have sensed previously.

"Expanding the right of workers to refuse hazardous work gives more power to safety-conscious workers because it permits them to draw the attention of both the union and the management to their specific complaints. . . . If individual workers acting on their own can legally refuse hazardous work without threat of job loss, safety-conscious workers can pressure their union to pay more attention to health and safety issues. It is important that this right not be collective; if it can only be exercised with the consent of the union it will have little impact on union bargaining priorities."[12]

Among union contracts covering over 1,000 workers, the Bureau of Labor Statistics found that about one-fifth of the 1,724 contracts studied gave employees the right to refuse hazardous work.[13] One such clause, from a Lockheed-Machinists union contract, simply states: "No employee shall be discharged or otherwise disciplined for refusing to work on a job not made reasonably safe or sanitary for him or that might unduly endanger his health."

Rights to Information

The labor law route and the OSHA route are also highlighted in disputes about information. Unions have asked employers to inform workers and the union what chemicals are being used in the work place. Also, unions have asked for employee exposure data and medical records, so that union-hired epidemiologists can analyze patterns of disease.

In three such situations, the companies refused to reveal the information, saying that it would disclose trade secrets, violate confidentiality of medical records, and would be burdensome to the company.[14] The National Labor Relations Board, which administers the labor law, has heard arguments in the three cases, but had not decided (as of August, 1981) whether *labor* law compels such disclosure to the employees' bargaining agent.

In 1980, the Occupational Safety and Health Administration issued regulations and draft regulations addressing information disclosure. Medical and exposure records are to be made available for examination and copying. Since this is a matter of OSHA—not labor—law, the right to information is not provided through union agency. The individual worker does not need a union to obtain this protection. In the final weeks of the Carter Administration, OSHA also issued a draft regulation regarding labeling of chemicals used in the work place. However, it was rescinded by the Reagan Administration.

A general situation may develop in which workers may reason that, in order to "find out"—perhaps "be sure"—that health standards are being observed, the union is their best route. By issuing standards, OSHA was giving individuals *an individual right,* flowing directly from the government. This may appear to displace unions, or obviate their fullest role as employee representatives. Insofar as OSHA avoids directly empowering individual employees, it creates some new avenues for union expression and appeal and representation.

Moreover, inspecting and enforcing standards is in a related two-choice situation. The less enforcement flows from "the government," the more attractive a union agent becomes. If the authorities cannot inspect and hold employers to a standard, the union can. It is *in* the plant (or wants to be); its members are the plant's work force. More than anyone else, members are interested in working conditions.

Union Responsibilities

The union movement appears to be on the verge of adopting health protection as one of its roles. However, if workers look to unions for this service, they also hold unions responsible when it is not adequately performed. Unions are being charged, under state negligence or common law, with failing to ensure a safe worksite or failing to warn employees about a dangerous substance or condition. In one case, the union contributed a part of the employer's settlement costs, to settle the company's charge that the union was negligent; it should have demanded safety protections. These moves are imposing a legal responsibility faster than unions can cope with it.

[11]Larry Drapkin, "The Right to Refuse Hazardous Work After Whirlpool," *Industrial Relations Law Journal,* 1980, volume 4, Number 1.

[12]Bacow, p. 114.

[13]U.S. Department of Labor, Bureau of Labor Statistics, *Major Collective Bargaining Agreements: Safety and Health Provisions,* Bulletin 1425-16, 1976, p. 20.

[14]See Chapter 4 for a discussion of the issue of informing employees of their health conditions.

The reaction has been greater caution about the unions' institutional role. For example, the AFL-CIO Building and Construction Trades Department has suggested contract language specifically exempting unions from liability on health and safety. It makes health protection the "exclusive responsibility" of the company. The Department also suggests a clause under which companies would indemnify unions against any liability for safety.[15] Finally, the Department asked its member unions to lobby state legislatures for a change in workers' compensation laws to exclude unions from liability in the same way as employers are protected.[16]

According to the same report, the most prominent union spokesman on health issues, Anthony Mazzocchi, Vice President of the Oil, Chemical, and Atomic Workers, commented on these legal avoidance tactics: "There will always be lawsuits. We have a recognized duty on safety and health. The best way to defend ourselves is to do what we're supposed to do."

Most of the issues surrounding union involvement in occupational health are present in this news "event" of early 1981:

"Burlington, N.J., Feb. 27—All 351 members of a

union local have struck a Hooker Chemical and Plastic Corporation plant here after learning that the company had brought a highly toxic gas, nitric oxide, into the plant without consulting the union.

'They snuck it in,' said William Boyea, who was among those picketing today at the plant, on the Delaware River. 'They're trying to kill us. They have ignored a request for safety and health protections. . . . ' "

"Robert Fine, director of Engineering and Maintenance and Environmental Safety for the plant, said in an interview that doctors for the union had recently asked for and been given a list of the chemicals and pigments employees worked with . . . .

"Richard Engler, associate director of the Philadelphia Area Project on Occupational Safety and Health, which helps unions identify and control on-the-job safety and health problems, said later that the information was being used to look into causes for deaths and illnesses of Hooker workers.

"He said, and Mr. Fine confirmed, that the Federal Occupational Safety and Health Administration last April cited Hooker for exposing workers here to excessive levels of vinyl chloride in the air in the plant over a period of years. Mr. Fine said that had since been corrected. Workers in the vat rooms wear respirators."[17]

[15] *Engineering News Record,* New York: © McGraw-Hill, Inc., July 31, 1980, p. 68.

[16] *Business Week,* August 4, 1980, p. 19.

[17] *The New York Times,* February 28, 1981, p. 26.

Appendix A
A Manufacturing Company's Mortality Surveillance Program

BACKGROUND:

This company initiated a mortality surveillance program in 1976 as part of its overall efforts to identify and control health hazards in the work environment. Initially, the program focused only on those deaths which were known to the company through its group life insurance program. In 1979, the program was expanded to include all active employees as well as terminated employees with a minimum of ten years' company service. This program is carried out by the epidemiology division of the Department of Health, Safety and Environmental Affairs.

PURPOSE:

The purpose of the mortality surveillance program is to alert the corporation to potential health hazards in the work environment as early as possible so that the appropriate action can be taken. Using the underlying cause of death as an indication of disease occurrence in the work force, this program seeks to identify excesses of particular diseases in the population that may have been caused by exposures to specific agents present in the work environment.

METHODS:

Notification of Deaths:

The epidemiology division is routinely notified of all deaths occurring among active employees and among retired employees covered by group life insurance. A listing of social security numbers of terminated employees with a minimum of ten years' company service is produced annually from company records and sent to the Social Security Administration for a determination of vital status of those employees not covered by group insurance.

Collection of Death Certificates:

For those employees covered by group insurance, death certificates are available through the Corporate Benefits Department, having been submitted as a proof of death. Where death certificates are not available from the Corporate Benefits Department, they are requested from the appropriate State Department of Vital Statistics.

Coding of Underlying and Contributory Causes of Death:

Coding of underlying and contributory causes of death has been performed according to the rules of the Eighth Revision of the International Classification of Diseases for all deaths occurring between January 1, 1970 and December 31, 1979. Starting in 1980, classification will be performed using the Accredited Classification of Medical Entities (ACME) program and the 9th Revision of the International Classification of Diseases.

Storage of Data:

A computerized "death file" containing an abstract of each coded death certificate is maintained as a confidential file with access limited to epidemiology staff. Hardback copies of all death certificates are kept in the epidemiology unit in locked file cabinets.

Analyses:

Proportional Mortality Ratios (PMR's) are calculated yearly for the entire corporation, for each of the major divisions, and for each plant within each division. PMR's are adjusted for age, sex and year of death and are standardized according to national, state and regional mortality statistics, when appropriate. For all PMR's where the observed and expected number of deaths exceed five (5), tests of statistical significance are performed using an adjusted Chi-Square test.

Protection of Confidentiality:

Results of PMR analyses are prepared for internal discussion and review at least annually. All analyses and summaries are based on grouped data only: in no instance is individual data released, discussed or presented in a manner that would violate an individual's right to confidentiality.

Appendix B
Corporate Epidemiology Policy of a Chemical Company

PURPOSE AND BACKGROUND

The Occupational Health Policy commits the corporation to identify occupational health risks associated with the work environment. The purpose of this policy is to describe how this commitment will be addressed through the corporate epidemiology program.

An important test of health effects from exposure to chemical and physical stress in the occupational environment is the demonstration of disease or lack of it in employed populations. While animal and in-vitro experimentation may serve as methods of screening various substances for toxic or biological effects, these results alone cannot be considered conclusive evidence for man. Properly conducted epidemiologic investigations of human populations are necessary to assist in the evaluation of human responses to specific occupational environments.

POLICY

1. The components are to obtain formal review and written approval from the Corporate Medical Department of proposals for all epidemiological investigations involving employees prior to any agreement for participation in the investigation. This includes proposals for internal studies that components plan to initiate and conduct, and proposals from trade associations, universities, unions, governmental agencies, or multicompany sponsors. Special attention will be given to the evaluation of the need for such a study and the quality and estimated cost of the proposal.

2. The components are to involve the Corporate Medical Department in monitoring the progress and conduct of all epidemiologic research involving employees to assure the quality of the research study.

SCOPE

This policy applies to all parts of the company in the United States and overseas.

PRACTICE

1. The Corporate Medical Department has established, maintains and operates a comprehensive population-based surveillance program for all active and retired employees. Vested employees are included, if possible. In addition, the company will conduct or sponsor through outside agencies, in-depth morbidity or mortality studies as indicated based on the results of the corporate surveillance program or other studies reported in the medical literature.

2. The Corporate Medical Department is staffed to provide components with professional epidemiological consultation and/or service as required or requested.

3. The Corporate Medical Department keeps abreast of new developments in occupational epidemiologic procedures and approaches, particularly in the area of morbidity studies in general and of reproductive and teratogenic studies in particular. The Corporate Medical Department routinely evaluates the results of reported epidemiologic studies of company populations or those of similar industrial populations and advises corporate and component management of the significance of the study results.

DELEGATION

1. Implementation

The component president and reporting line management have the responsibility for the occupational health of the component's employees, and as such, for the overall implementation of this policy. It is the responsibility of plant physicians, nurses, industrial hygienists, employee relations managers, and health managers to inform the Corporate Medical Department of any evidence suggestive of the need for an in-depth epidemiology study and to cooperate in providing the necessary information for any epidemiological activity.

2. Auditing

The Corporate Health, Safety & Environmental Affairs Department has the responsibility to audit component line performance against this policy.

3. Policy Development and Modification

The Corporate Health, Safety & Environmental Affairs Department has the responsibility for initiating modifications to this policy as appropriate.

Code of Ethical Conduct for Physicians Providing Occupational Medical Services

These principles are intended to aid physicians in maintaining ethical conduct in providing occupational medical service. They are standards to guide physicians in their relationships with the individuals they serve, with employers and workers' representatives, with colleagues in the health profession, and with the public.

Physicians should:

1. accord highest priority to the health and safety of the individual in the workplace;

2. practice on a scientific basis with objectivity and integrity;

3. make or endorse only statements which reflect their observations or honest opinion;

4. actively oppose and strive to correct unethical conduct in relation to occupational health service;

5. avoid allowing their medical judgment to be influenced by any conflict of interest;

6. strive conscientiously to become familiar with the medical fitness requirements, the environment and the hazards of the work done by those they serve, and with the health and safety aspects of the products and operations involved;

7. treat as confidential whatever is learned about individuals served, releasing information only when required by law or by over-riding public health considerations, or to other physicians at the request of the individual according to traditional medical ethical practice; and should recognize that employers are entitled to counsel about the medical fitness of individuals in relation to work, but are not entitled to diagnoses or details of a specific nature;

8. strive continually to improve medical knowledge, and should communicate information about health hazards in timely and effective fashion to individuals or groups potentially affected, and make appropriate reports to the scientific community;

9. communicate understandably to those they serve any significant observations about their health, recommending further study, counsel or treatment when indicated;

10. seek consultation concerning the individual or the workplace whenever indicted;

11. cooperate with governmental health personnel and agencies, and foster and maintain sound ethical relationships with other members of the health professions; and

12. avoid solicitation of the use of their services by making claims, offering testimonials, or implying results which may not be achieved; but they may appropriately advise colleagues and others of services available.

Adopted by the Board of Directors
of the American Occupational Medical Association, July 23, 1976
Adopted by the Board of Directors
of the American Academy of Occupational Medicine, January 15, 1977

Scope of Occupational Health Programs and Occupational Medical Practice

Committee Report

This statement was prepared by the Occupational Medical Practice Committee of the American Occupational Medical Association: Bruce W. Karrh, M.D., Chairman; Bruce H. Bennett, M.D.; Caesar Briefer, M.D.; Robert M. DeuPree, M.D.; William G. Mays, M.D.; James W. Mitchell, M.D.; Robert H. Moore, M.D.; Billie H. Shevick, M.D.; W. Lloyd Wright, M.D. The statement was approved by the AOMA Board of Directors at its meeting April 20-21, 1979, in Anaheim, Calif.

Introduction

Occupational medicine as a specialty and the practice of occupational medicine have undergone many changes in recent years. These have been caused largely by changing expectations of society in general, and employed persons in particular, as exemplified by legislation directed toward providing a safe and healthful workplace, such as the Occupational Safety and Health Act and the Toxic Substances Control Act. There has also developed great concern over the chronic health effects of long-term exposures to low levels of chemical and physical agents. The result has been to increase the demands on the time of occupational physicians while requiring that they be more knowledgeable and experienced in both clinical and occupational medicine, toxicology, epidemiology, industrial hygiene, and administration.

While there has been a change in the demands on physicians practicing occupational medicine, the scope of occupational medical practice has also enlarged, resulting in an increase in the components and services of an occupational health program.

An occupational medical program must, as a minimum
- obey all relevant laws and regulations,
- take all feasible steps to keep industrial operations and/or products from having an adverse effect on the health of employees, customers or the public, and
- accept responsibility and assist in the provision of necessary care in cases where health is harmed due to industrial operations and/or products.

The specific contents of an occupational health program will be largely determined by the functions of the work organization and the specific workplace activities and potential hazards present.

Comprehensive occupational health programs require the skills of persons trained or experienced in a variety of disciplines, including clinical and occupational medicine, industrial hygiene, toxicology, epidemiology and biometry, occupational health nursing, safety engineering, and human factors engineering. Such a program requires an organizational structure that insures communication and coordination of activities between these various, seemingly diverse, but interdependent, skills. The exact organizational relationship necessary to fulfill the various responsibilities varies between companies and can only be generalized in this document.

The organization must report at a management level high enough to assure that top company management is fully informed of all significant occupational health activities, problems and concerns so that appropriate action can be taken where necessary to assure a safe and healthful workplace.

An occupational health program should not be used as a means to further the specific interests of management or unions but to provide methods and procedures to assist in providing a safe and healthful workplace for all concerned.

Records which are developed for retaining the data from an occupational health program—medical data on employees, exposure data on employees or groups of employees, work assignment data, toxicology and industrial hygiene data, and epidemiological data—must be kept in such a way as to insure the degree of confidentiality the data require. Procedures for preserving this confidentiality, yet allowing access to those with a bona fide need to know, must be developed.

Employees who are the subject of medical or exposure records should be informed of the existence and contents of these records. Likewise, employees should be informed of the appropriate toxicologic and epidemiologic data which are pertinent to their potential workplace exposures.

This document will list those components and services which are considered a necessary or essential part of all occupational health programs and those which are desirable if time and availability of medical and paramedical personnel permit. These latter are elective components of an occupational health program.

Occupational Health Program

A. *Essential Components*
  1. *Health Evaluation of Employees*
     Employees should be fully informed of the results of each health evaluation, whether or not abnormalities are detected. When abnormalities, or questionable abnormalities, are detected, the employee should be informed and referred to the personal physician or appropriate referral physician for further diagnostic evaluation or treatment.
     Evaluations should be carried out on the following occasions:

     (a) *Pre-assignment*—Health status, including assessment of emotional status, should be assessed prior to making recommendations to management regarding the assignment of an employee to a job to assure that the person can perform the job safely and efficiently without endangering the person's safety or health and that of others. This recommendation shall be based on any or all of the following:

        (1) Medical history
        (2) Occupational history
        (3) Assessment of the organ systems likely to be affected by the assignment
        (4) Evaluation of the description and demands of the job to which assignment is being considered.

     (b) *Periodic*—The health status of the employee should be periodically reviewed where there is a likelihood that workplace exposures or activities could have an adverse health effect. This review may be limited to those organs or systems most likely to be affected.

     (c) *Post-illness or injury*—The health status of an employee should be reevaluated following absence from work due to illness or injury to assure that the individual has sufficiently recovered from the illness or injury to perform the job without undue risk of adverse health or safety effects to the individual or others, and that the employee is not taking any medication which increases the risk of illness or injury due to the workplace.

     (d) *Termination and Retirement*—The health status of an employee should be evaluated at the time of termination or retirement. The employee should be informed concerning his or her health status and advised of any adverse health effects due to the job.

  2. *Diagnosis and treatment of occupational injuries or illnesses, including rehabilitation.*
     Occupational injuries and illnesses should be diagnosed as promptly as practical and treated as appropriate within the capabilities of the workplace medical facility. The occupational health personnel for a workplace are uniquely qualified to diagnose occupational illnesses and injuries because of their knowledge of the workplace. The occupational physician should also be knowledgeable regarding rehabilitation programs and facilities in the area. Frequently the workplace can be used for rehabilitating employees, especially where selective work can be provided.

  3. *Emergency treatment of nonoccupational injury or illness.*
     The occupational health program should provide treatment for emergency conditions, including emotional crises, which occur among employees while at work. This treatment may only be palliative and to prevent loss of life and limb or, where personnel and facilities are available, may be more definitive.

  4. *Education of employees in potential occupational hazards which may be specific to the job, instruction on methods of prevention and on recognition of possible adverse health effects.*
     Every employee should know the potential hazards involved in each job to which he or she is likely to be assigned. This instruction must include methods of recognizing and preventing possible adverse health and safety effects from the workplace. The employee must be instructed to report any adverse health effect to his or her supervision and to the occupational medical personnel.

  5. *Evaluation of programs for the use of indicated personal protective devices—ear plugs, safety spectacles, respirators, etc.*
     The occupational health personnel should develop techniques and expertise to assist management in properly fitting personal protective devices, determining that the devices provide adequate protection to employees, and educating the employees in proper utilization and care of the equipment.

6. *Assist management in providing a safe and healthful work environment.*
Occupational health personnel should periodically inspect and evaluate the workplace, looking for potential health and safety hazards. Management should be informed when such hazards are found and, where expertise exists, make recommendations for abatement of the hazard.

7. *Toxicological studies on chemical substances which have not had adequate toxicological testing.*
Occupational health personnel should advise management on the adequacy and significance of toxicological test data pertinent to the workplace. Where adequate data do not exist, occupational health personnel should recommend appropriate resources for testing.

8. *Biostatistics and epidemiology when adequate data are available and a need exists to evaluate the experience of persons at risk.*
The occupational health program should assure that data on employee work experiences and exposures and medical occurrences are accumulated and retained. When appropriate, these data should be used to conduct epidemiological studies to assess the effects the workplace may have had or is having on the employees.

9. *Maintenance of occupational medical records.*
The occupational health program must maintain occupational medical records on each employee, documenting the reasons for and the results of all physical examinations and visits to the medical facility. Ideally, these records should contain data sufficient to reproduce a chronology of the employee's medical occurrences, illnesses, and injuries.
These data must be maintained confidentially. Procedures for preserving this confidentiality, yet allowing access to those with a bona fide need to know, must be developed.

10. *Immunization against possible occupational infection.*
Where employees work jobs with potential exposures for which there is an effective immunization, this protection should be provided to the employee.

11. *Participate with management in planning, providing, and assessing the quality of employee benefits.*
The occupational health personnel are best qualified to assist management in evaluating employee health benefits and the costs of such benefits.

12. *Assist in interpretation and/or development of governmental health and safety regulations.*
Occupational health personnel are uniquely qualified to assist in interpreting and developing these regulations and to assure they effectively provide the necessary protection to the employee in a manner that is most cost/effective and best utilizes professional occupational safety and health personnel.

13. *Periodic evaluation of the occupational health program.*
This is necessary to assure the program meets its objective effectively. The mechanism for this assessment will vary, but may exist in the program itself.

B. Elective Components of Occupational Health Programs
1. *Palliative treatment of disorders to allow completion of workshift or for conditions for which an employee may not ordinarily consult a physician.*
Personal medical care may be provided to employees where suitable medical care is not available in the community. This may include early diagnosis, definitive treatment and follow-up but only under certain appropriate and limited conditions.

2. *Repetitive treatment of nonoccupational conditions prescribed and monitored by personal physician (physiotherapy, routine injections, etc.) or if the employee's personal physician approves this approach.*
This provides a service to the employee, the personal physician and the employer by minimizing the time the employee must be off the job, the expense to the employee and the personal physician's time.

3. *Assistance in rehabilitation of alcohol or drug dependent employees or those with emotional disorders.*
The workplace provides a unique opportunity to provide this type of assistance to employees and their families.

4. *Health education and counseling (for example, mental health, hypertension control, diabetes control, obesity, physical fitness, smoking cessation programs, etc.).*
The workplace provides a unique opportunity for effective health education and health maintenance programs. These may include health counseling for employees and their family members. Occupational health personnel should assist employees by offering advice and counsel regarding the personal health care the employee or family member receives from the local medical community.

5. *Assist management in control of illness-related absence from job.*
Occupational health personnel are uniquely qualified to assist management in assessing the reasons for an employee's poor performance or absence from work due to illness or injury and in determining when they are well enough to return to work safely.

6. *Disaster preparedness planning for the workplace and, when appropriate, the community.*
Occupational health personnel should assist management in preparing a plan for disaster preparedness. Since community facilities and health and safety personnel are such an essential part of dealing with an emergency at the workplace, such planning should be done in conjunction with that of the local community.

7. *Immunization against nonoccupational infectious diseases.*
Frequently the workplace is equipped to provide appropriate immunizations for international travellers and for other nonoccupational conditions.

Summary

The goal of an occupational health program is to insure a safe and healthful workplace. The role of the occupational health professional is to assist in providing this safe and healthful workplace. Although this document is to provide a compilation of the necessary and important parts of an occupational health program, it does not detail how these programs should be structured or how each will function. More assistance on the details can be obtained from suitable reference materials, such as those on the attached brief list, or by consultation with persons trained or experienced in this field.

Reprinted from Journal of Occupational Medicine
July, 1979, Volume 21, No. 7,
pp. 497-499.

Appendix E

Special Report

Epidemiological Requirements for Medical-Environmental Data Management

Sidney Pell, Ph.D.

A medical information system that is to be used as a source of data for epidemiologic research requires more data than are routinely generated in occupational health programs. Moreover, the quality of the data must conform to the standards required for research purposes.

To meet these requirements, the system must supplement existing data by including other sources of medical information, demographic characteristics associated with the risk of disease, company population census statistics, and comprehensive work histories. Furthermore, to insure the completeness and reliability of the data, the system should provide for case finding mechanisms, validation of diagnoses and cause of death statements, follow-up of terminated employees, and data validation procedures.

If a data collection system is to be a satisfactory source of data for epidemiological studies, the information that goes into the system must meet the rigid requirements of epidemiological research. The mere mechanical transfer of whatever data are available from their source to a computer is grossly inadequate because epidemiologic studies require more information than what becomes available routinely, and, equally important, the information must have greater precision, reliability and completeness than would be acceptable when that information is used for other purposes.

The data required for epidemiological studies of occupational disease fall into three general categories: medical, work history, and demographic. In an industrial organization, one would expect to get the medical data from records generated by the company's medical program, and the work history and demographic data from payroll and personnel records. How adequate are these data sources for epidemiologic research?

Medical History

Periodic Health Examinations. — Occupational medical programs vary with respect to their scope and com-

prehensiveness.[1] Health examinations may be given either at regular intervals, only at the time of employment, or at some other designated frequency, depending upon the employee's age or occupational status. The examination may vary from one company to another and may even vary among the employees within a given company. Moreover, physicians within a company may differ considerably with respect to the thoroughness with which they give the examination, thereby affecting the quality of the medical data. And finally there may or may not be adequate follow-up of positive findings to determine whether those findings led to diagnoses of specific diseases. If a positive finding is not followed up by additional tests, or by contact with the employee's personal physician, the occurrence of a newly diagnosed disease may not become known to company medical personnel and, therefore, not be recorded in the medical records.

Sickness absences, dispensary visits, and consultations are other types of information that may appear in medical records and contribute to the total picture of an employee's medical history, but their value for epidemiologic research depends upon the reliability of the diagnoses and to what extent leads have been followed up to get all pertinent information.

Thus, there may be several gaps in data generated by company medical programs that impair their usefulness for epidemiological studies. If, for example, a study is undertaken to compare the prevalence of certain chronic diseases in a given company with that of other populations, or to compare prevalence rates among sub-groups within the company, it is essential that every case in the company be diagnosed and entered in the medical records.

In 1954, a study of chronic disease in the Du Pont Company was conducted by the University of Pittsburgh, utilizing the company's medical records.[2] The study revealed great variations in prevalence rates among plants. The investigators did not believe the differences were real, but suggested that they probably reflected "differences in examining procedures in the different plants, . . . differences in recording procedures, and variations in completeness of reporting to the central office."

Periodic health examinations in occupational health programs are intended primarily to screen employees for early detection of chronic disease. They are not intended to identify and provide a definitive diagnosis of every disease that occurs in the worker

From the Medical Division, E. I. du Pont de Nemours and Company, Wilmington, Delaware.

Presented at the American Occupational Medical Association annual meeting, April 11-14, 1978, New Orleans, La.

population. Thus, if a data system is to be useful for epidemiologic research it requires additional information to supplement the data that are provided routinely by an occupational health program.

Insurance Records. — Potentially important sources of medical data that may supplement the data made available by periodic health examinations are company sponsored sickness and life insurance plans. Claims filed by the employees or their beneficiaries usually require supporting documents that contain information pertaining to the employees' medical history. For example, an accident and sickness plan may require a statement of the diagnosis by the attending physician, and a life insurance plan may require a death certificate to support a claim made by the beneficiary.

It is important to keep in mind that diagnostic statements and cause of death information should not be accepted uncritically and mechanically put into the data system. These statements must be carefully reviewed by appropriately trained personnel to detect errors, missing information, and inconsistency with other sources of information. Follow-up must be undertaken to make whatever corrections are necessary before the diagnoses and causes of deaths go into the data system.

If the incidence or prevalence of disease is to be investigated in epidemiological studies, it is important, therefore, that all potential sources of medical information be utilized to develop case finding mechanisms, and that diagnostic criteria be established to insure the validity of the data.

Work Histories

Information on work history can be obtained from payroll and personnel records. These records usually show job titles and areas in the plant where the employee was assigned, including dates in and out of jobs and work areas. In some instances, these types of information may be sufficient to identify workers who have been exposed to a specific substance, or they may be useful in an initial phase of a study. Most of the time, however, this kind of information is too general and imprecise. Job titles, such as "mechanic," "operator," "laborer," and "machinist" reveal very little. The areas and manufacturing processes to which the workers were assigned may encompass a large variety of tasks involving many substances.

For epidemiological studies, it is important not only to identify all workers who have been exposed to a specific substance, but to obtain the level and duration of their exposure. Thus, the data system must also include the results of industrial hygiene surveys and environmental monitoring data, as well as the kinds of personal protection worn by the workers.

Recording all the information that is pertinent to exposure is a formidable task. It is made especially difficult by the many substances to which a single worker may be exposed during his working lifetime, by frequent changes in job assignments, by the variability of the environmental concentrations of substances in the workplace, by the mobility of the worker during the course of a working day, and by changes in manufacturing processes.

How these problems can be dealt with is beyond the scope of this article, but it should be mentioned that, whatever system is developed, it should be one that has the capability of identifying all workers that have been exposed to a specific substance and of providing sufficient information on level and duration of exposure to develop a dose-response curve, if the substance does have an adverse effect on health.

Demographic Information

Epidemiological studies must include all types of demographic characteristics that are associated with the risk of disease so that they can be taken into account in the analysis of the data. In addition to such obvious disease related characteristics as age and sex, the data system should include race, socioeconomic status, and family composition.

Race. — Total mortality, age-adjusted, is about 45% greater among nonwhites than whites.[3] The major diseases that contribute to this excess are cerebrovascular disease, hypertension, and certain types of cancer, such as cancer of the lung, prostate, esophagus, pancreas, and stomach.[4] Whites have higher cancer incidence rates for such sites as the colon, rectum, bladder, leukemia, and lymphomas.

Socioeconomic Status. — Socioeconomic status tends to be inversely related to mortality,[5][6] to the prevalence of illness and sickness disability among employed persons,[7] and to the incidence of certain types of cancer, such as cancers of the lung, esophagus, stomach, buccal cavity, and uterine cervix.[8]

The primary indicators of socioeconomic status are income, educational attainment, and occupation. Since this kind of information is usually recorded in payroll or personnel records, it should be readily accessible for entry into an epidemiological data system.

Reproduction History. — Recently, there has been increased concern over the potential effects of some chemical compounds on human reproduction. Some chemicals are known or suspected to be mutagens, teratogens, embryotoxins, or to impair fertility in some other way. For this reason, it would be advisable to incorporate into an epidemiological data system information needed to conduct fertility studies, such as marital history, the birth date of the employee's spouse, the number of children and the birth date of each child. Occurrences of spontaneous abortions, fetal deaths, and perinatal deaths are also needed for these studies, but the feasibility of getting this information routinely and into a data system is open to question and needs to be explored.

Company Population Statistics. — Demographic information is also needed to provide periodic census data for the company population. At least annually, population statistics should be obtained which show the number of company employees, broken down by age, sex, and race. These statistics should be obtained for the company as a whole and for the company's plants, offices, laboratories, and other installations. They are needed to provide the denominators for the computation of age-sex-race specific incidence or prevalence rates in morbidity studies and death rates for mortality studies. These statistics are also needed to compute rates that are to be standardized for age, sex, and race.

Identification Number. — The data system should also include an identification number for each employee, preferably the social security number. If some other number is used, such as a payroll number or any other number assigned within the company, that number must, like the social security number, be unique for that person. After the employee is terminated, it must not be reassigned to a new employee. Since the number will be used to link computer files and to search computer files to gather data for long-term studies, reassignment of identification numbers will obviously lead to errors in assembling medical and work histories for individual employees.

Retirees and Other Terminated Employees

In occupational health research, it is essential that the studies include workers who have retired, resigned, were laid off, or whose employment has been terminated for any other reason. If some facet of the employee's job has an adverse effect on his or her health, the effect may not become evident until several years after exposure and when the worker is no longer employed, or is employed elsewhere. The latent period following exposure to a carcinogen is a well known illustration of this problem.

If health effects are investigated by a mortality study, one must keep in mind that if, as a result of a serious illness, an employee becomes too disabled to work, he or she may be laid off or given an early retirement pension before the usual mandatory retirement age of 65 years. If the illness is ultimately fatal, death may not occur until after employment termination and would, therefore, not be included in the data if the study were confined to active employees. This problem will occur whether or not the illness is job related. In a mortality study of Du Pont Company employees and retirees, we found that the average annual death rate per 100,000 among male active employees was 339.3.[9] All the men in this category were below the age of 65 because retirement is mandatory at that age. When we added to that population retirees below the age of 65, the death rate increased to 458.2. The death rate of the retired population alone was very high, i.e., 5,772.2, because this group of retirees was composed chiefly of employees who had been given disability pensions as a result of serious illnesses, many of which had a fatal outcome.

If mortality statistics of retirees are to be combined with those of active employees, it is necessary, as in the case of active employees, to get periodic population statistics of the company's retired population, broken down into the same categories mentioned earlier.

It is especially important to get data on terminated employees in a prospective cohort study. In this type of study, a group, or cohort, of employees exposed to a suspected work hazard is followed over a certain period of time to determine whether its morbidity or mortality experience is in any way significantly greater than that of a comparison population. During the course of the study period, a portion of the cohort will be lost to observation because of employment termination. If the experience of this group is subsequently different from the group that remains under observation, the results of the study would be biased.

Getting information on terminated employees may be difficult. In mortality studies, employees who retire on a company pension can easily be followed by using records maintained to administer the pension plan. These records can provide the data system with the vital status of the retirees and the date of death. The cause of death could be obtained if retirees are covered by the company's life insurance plan and a death certificate is required to support the claim made by the beneficiary. Otherwise, the death certificate would have to be obtained from the state health department.

Morbidity data on retired employees usually cannot be obtained routinely, if at all, because after retirement they may no longer participate in the company's medical program or be covered by an accident and sickness plan. Morbidity information on a portion of a company's retired population may become available by informal contacts or by health examinations provided by the company to those retirees who wish to take them. These data must be used cautiously in epidemiological studies because they may be biased by significant differences in disease incidence between retirees who remain under medical surveillance and those who do not. Moreover, the quality of this medical information may be well below that obtained among active employees.

Employees who leave a company before they attain eligibility for retirement benefits cannot be traced through company records, and therefore present a special problem in cohort studies. Mortality information can be obtained for these persons by obtaining their vital status and date and place of death from the Social Security Administration, and then getting death certificates from state health departments. Other sources for follow-up include the Internal Revenue Service, city directories, and forwarding addresses from the U.S. Postal Service. Getting this kind of information routinely on all such terminated employees to feed into a data system is expensive, time consuming, and not essential to the success of an epidemiological program. This type of data should be obtained, however, for employees who are lost to follow-up in specific cohort studies.

Summary

In establishing a system to provide data for epidemiological research in occupational health, it should be recognized that such studies require more information than what usually becomes available routinely in occupational health programs. Additional types of data that may have to be put into the system include supplementary diagnostic information, level and duration of exposure to potentially hazardous substances, demographic information related to socioeconomic status and fertility, population census statistics for the entire company and various company installations, and follow-up data on retired employees. The quality of the data is especially important for epidemiologic research. Therefore, data validation procedures need to be established whereby the information that goes into the system is reviewed for precision, completeness, and reliability.

References

1. Forbes JD et al: Utilization of medical information systems in American occupational medicine — A committee report. *J Occup Med* **19**:819-830, 1977.

2. Densen PM, D'Alonzo CA, and Munn MG: Opportunities and problems in the study of chronic disease in industry. *J Chronic Dis* **1**:231-252, 1955.

3. U.S. Public Health Service: Vital Statistics of the United States, 1972. Vol. II, Mortality, Part A. Rockville, 1976, pp 1-6.

4. Levin DL et al: Cancer Rates and Risks, Ed. 2. National Cancer Institute, 1974, pp 15-17.

5. Guralnick L: Mortality by Occupation and Industry Among Men 20 to 64 Years of Age: United States, 1950. Vital Statistics - Special Reports, Vol. 53, No. 2, Sept. 1962.

6. Nagi MH and Stockwell EG: Socioeconomic differentials in mortality by cause of death. *Health Serv Reports* **88**:449-456, 1973.

7. Gleeson GA: Selected Health Characteristics by Occupation, United States, July 1961 - June 1963. Vital and Health Statistics, Series 10, No. 21. Washington, D.C., 1965.

8. Levin DL et al: Cancer Rates and Risks, Ed. 2. National Cancer Institute, 1974, pp 58-61.

9. Pell S: The effects of selection factors on epidemiologic research in employed populations, in The Nature of Biostatistics. University of Pittsburgh, 1969, pp 175-194.

Reprinted from Journal of Occupational Medicine
August 1978, Volume 20, No. 8
pp. 554-556
©JOM 1978

Appendix F
The Organizational Environment and Ethical Conduct in Occupational Medicine: A Perspective*

Robert W. Ackerman, D.B.A.
Vice President, Preco Corporation
West Springfield, Massachusetts

M Y FOCUS will be primarily on large corporations, not because problems do not exist in small ones, but because a large corporation presents the most difficult problems to deal with from an organizational standpoint. They also have the resources to innovate, and tend to set the ethical tone for other employers in this country.

To approach the matter of ethics in the corporation I shall have to go a bit beyond occupational medicine to the problems associated with corporate response to social change. Finally, I shall indicate some of the changes I believe necessary to make corporations more responsive to social forces.

The background for this talk is a research project initiated by Raymond Bauer at the Harvard Business School in progress for the last five years. More than four dozen large corporations have been studied in depth as to how managers approach social issues, how decisions are made, and the outcome of those decisions. We have tried to draw together our findings and conclusions in a more general framework to help corporations to be more responsive to the social forces around them.

It should be apparent that since the early 1960s social demands on the corporation have grown enormously. The occupational health and safety issue is certainly a major example of the increasing social pressure on corporate America. Other issues have involved the environment, equality in employment, consumer protection, and so on. The petitions for reform that accompany such issues have a number of characteristics in common.

First, they involve higher order consequences from business decisions formerly made on narrowly technical and economic grounds. Second, they arouse constituencies which had either not existed before or had been weak and disorganized. Such constituencies often have popular appeal because, typically, matters of justice and equity are raised in which the powerful corporation is pictured as menacing the environment, subjecting its employees and customers to unnecessary and unexplained risks, etc. Third, they involve technical uncertainty— problems without known solutions—and this uncertainty in turn poses questions about the nature and magnitude of the resources to be marshalled by the corporation in its response. Finally, they involve continually changing public expectations and government regulations because what was condoned yesterday may be protested today and forbidden tomorrow.

The corporation has not always responded particularly well to these challenges, and although I expect that corporations are more sensitive and responsive to social issues than five or six years ago, the public's perception of this fact is not very clear. An annual Harris poll, wherein people are asked about their confidence in various institutions, reflects a drop for the large corporation from 55% in 1966 to 15% in 1976. People have also expressed diminishing confidence in other institutions in our society, but nevertheless the corporation has reason to be concerned about its popular image.

Two reasons, which I believe woefully inadequate, are frequently offered for the corporation's failure to address the demands of the public. The first is that managers are unethical people motivated by greed and the quest for power. Certainly there are bigots and charlatans in large corporations as there are in other spheres of public and private life, but I doubt that this is a good general explanation for the unethical behavior ascribed to the large corporation. In fact, I would argue that the organizational structure in the corporation today breeds an official amorality. The second reason for failures in corporate responsiveness is typically laid to the costs involved. There are certainly cases in which companies and even industries cannot afford to respond to pressures brought to bear on them. Yet there is little evidence of otherwise healthy businesses succumbing to rising social costs which cannot be transmitted to the public through higher prices. Indeed, we as a nation have increasingly shown a willingness to adjust our expectations as social costs escalate. Unresponsiveness is far more likely to result from strangulation by bureaucratic red tape if the whims of government officials seeking to implement the public's wishes are not satisfied.

*Presented at a *Conference on Ethical Issues in Occupational Medicine* cosponsored by the New York Academy of Medicine and the National Institute for Occupational Safety and Health and held at the Academy June 21 and 22, 1977.

I believe that a third reason for unresponsiveness is at the root of the corporations' problems, the pattern of organization that has evolved over the last three decades. Since World War II large American firms have significantly extended the geographical scope and diversity of their activities. To manage operations of increasing size and diversity, a divisional organizational structure evolved and became widely established. This organizational form, pioneered by duPont and General Motors in the 1920s, constitutes the dominant structure in American business. A recent study of the "Fortune 500" indicated that by 1969 approximately 80% were organized in this fashion, whereas in 1949 the number had been only 24%.

This organizational structure differs fundamentally from what such business critics as John Kenneth Galbraith would have us believe is typical of American industry for several reasons.

First, the individualized organization normally separates responsibility between managers at the headquarters or corporate level and those at the operating division or profit-center level in the organization. Typically, corporate managers are responsible for overall financial affairs and planning and for the policies and systems that control the corporation and measure the performance of individual managers, but not for the management of the component divisions of the corporation. The organization tends to be characterized by a "bottoms up" style of planning. Courses of action are developed at division levels, formed into plans and budgets, and reviewed up the line. Hence, while the firm may be huge in its entirety, it is far from monolithic, and the power of the division manager to influnce the course of the entity under his direction is considerable.

A second characteristic is the reliance on financial control systems. Short-term financial performance has become the common denominator through which the divisionalized organization is managed. It has a commanding influence on how resources are allocated among divisions and has become the center piece in determining rewards for individual managers. Career paths for middle managers are naturally shaped by the performance of their divisions as reported in these control systems.

One consequence of this pattern of organization is that communication between corporate executives and the managers who run the divisions has tended to become more abstract and oriented toward short-term plans and results. The social system in the large corporation through which the values of the chief executive were once transmitted to the organization is changed. I do not wish to overemphasize these characteristics, but they nonetheless suggest that the large corporation is quite different from its predecessor of 30 and 40 years ago.

The divisional organization structure has been extremely successful in managing traditional business problems, and it is probable that the United States would not have enjoyed the growth it had over the past several decades had it not been for this administrative innovation. However, social demands have posed difficulties for the divisionalized firm that are not easily overcome. Ironically, the strengths of the organization have been barriers to social responsiveness. Those interested in change are well advised to appraise the situation rather carefully.

First, social demands create difficulties in fixing responsibility to establish and to implement policy and even to discuss them among managers in the organization. The chief executives who by design are not involved in day-to-day operating problems find themselves challenged at annual meetings, by the press, and in personal conversations on very specific issues that relate to matters over which they have no direct control. They feel and are made to feel that they have final responsibility for the conduct of the business.

On the other hand, the chief executives are decidedly reluctant to usurp the responsibilities assigned to division managers, knowing that such interventions will reduce the division managers' accountability for results. Unfortunately, operating managers often lack the skills to handle social problems in their organizations and, without policy guidance from the top, find it difficult to make the tradeoffs that may be necessary between social responsiveness and short-term financial results. Consequently, if responsiveness costs money, they may not assume the responsibility.

Second, social responsiveness is not easily measured by the financial control system. While some of its costs may be determinable and immediate, benefits are usually unknown and often occur in the future. How, for example, should one handle the accounting for the installation of safety guards on a machine? The cost of the guards can be viewed as an expense (or an investment) but the benefits in terms of a lower probability of injury or higher worker satisfaction is conjectural. Thus, the performance incentive for an operating manager to take an aggressive posture is lacking. Indeed, there may seem to be risks in involvement with problems that are difficult to understand, that have uncertain outcomes, and that will take time away from the operations of the business. Why stir sleeping dogs?

Although I may have painted a gloomy picture, I do not think that pessimism is necessarily justified. Over the years, corporations have learned to respond to social demands. That is, they have been able to implement socially motivated policy at business or division levels that overcome the organizational barriers to change.

Three ingredients in this learning process become evident over a period that may extend for five to seven years. First, an awareness of an issue and a commitment to change gathers force at the top of the corporation and eventually evolves into policy directives. Without this commitment the possibility of change in advance of the time that is prescribed and effectively enforced by regulation is remote. The top officials in the large corporation must take an active role in a way that is really unnatural in the divisional organization, which is geared to shielding such managers from the details of operations.

The power of the policy should not be overestimated. For the most part, such statements will be ignored if they are treated as the solution all by themselves. A policy cannot possibly be specific enough to guide the operating manager to action, and cannot provide the skills or understanding necessary to find solutions. At best it is a thoughtful expression of intent, expounded by a manager not in a position to insure its implementation.

A second step or ingredient is the provision of technical skills, generally in the offices of a person I shall call the social issue specialist. This person is typically appointed at the right time, but often for the wrong reason. The chief executive may decide that if the problem can be segregated and assigned to people who understand the environment and the technical basis for responding to it, a satisfactory response will in some way or other be forthcoming.

The rationale for the appointment of a social issue specialist and the creation of a department charged with the responsibility to insure responsiveness is plain enough. The environment is murky, as are the solutions, and the organization itself is unprepared to act. External relations must be managed and are themselves evolving over time as public interest matures and is mirrored in government action. Methods for planning and coordinating corporate initiatives must be established. We are aware of social issue specialists in the health and safety area in our firms or organizations. Very often they are in the personnel area, but I have seen them in the environmental control area as well, depending on whether the basic occupational health and safety problem is one of human relations or of engineering.

Although social issue specialists play a vital role in corporate responsiveness, their failure rate is unfortunately high. Frequently, the initial incumbents in these jobs are either outsiders who know a great deal about the external forces involved but have comparatively little understanding of the organization that they have come to work with, or they are managers assigned from somewhere else in the organization without a background in the problem. In either case, the job may be viewed as out-of-the-normal career path and as lacking the kind of influence necessary to move solutions along.

These managers often are caught up in a web of relations, and have a great deal of responsibility and comparatively little authority. They must confront division managers who are extremely defensive about whatever will have an impact on the performance of their units and who are very suspicious of those on the corporate staff. The social issue specialist is also caught between operating managers and environmental groups with competing demands and interests. Should organizational conflicts develop which call for resolution by the chief executive, the social issue specialist will often find that he is a poor match for a division manager when it comes to debating his case. It is a difficult role to play and, by itself, will not produce aggressive social responsiveness by the corporation.

A third ingredient must be added to elicit change and to make division managers responsible for implementing policy. This final step is not easily taken, and often is not taken until forced by external pressure. The chief executive must be willing to clarify responsibilities and to make the response to social demands part of the operating manager's job. To create the necessary incentive in the traditional mold of the divisional organization, a part of this manager's performance evaluation should rest on the effectiveness of the response.

Several implications for this setup may be of particular interest. The first is in the nature of change itself in the large corporation. Appeals to ethics, to morality, or to reason alone are not enough. They touch neither the vital nerve that theologians or economists would have us believe is the guiding influence on corporate behavior, although both may have their place. Yet change in the large corporation is certainly possible. It is an adaptive entity and will respond to external pressure. To manage change, however, one must be sensitive to the structural barriers to be overcome, and in particular to the career system through which managers are rewarded for their efforts. The process of adaptation requires that attention be given to the three factors discussed above.

The first is the visible, tangible support of the chief executive. Policy statements on an issue evolve as more is learned, but the chief executive's commitment to grapple responsibly with the issue must in some way be sustained. Second, technical support must be provided, not just with respect to the social issue itself, but to the administration of the process for responding to it within the firm. The organization must be taught how to manage its response, not just through a doctor's eyes or a specialist's eyes, but through an operating manager's eyes. And, third, a place must be given to the issue in the performance-evaluation system. Operating managers must feel that successes and failures in implementing corporate social policy form a part of the basis upon which their careers will be determined.

Thus, drafting a policy or goal is the beginning rather than the end of the adaptive process. Moreover, the process itself has limitations. The chief executive has a limited amount of time and cannot be expected to treat all issues with the same degree of interest or as having equal importance. More important, the amount of change that can be imposed on the administrative system in the firm is limited. For instance, in one case involving equal employment it was finally decided that 15% of the manager's performance-review ought to be allotted to success in meeting affirmative-action targets. This procedure created problems, however, because questions were soon raised: What about environmental concerns? Or occupational safety and health? Or product safety? Should each get 15%, 10%, or 5%? How can a manager's job be understood if it is cut up into little pieces? The level of anxiety and frustration among middle managers was terrific. The problem is very difficult, and probably cannot be addressed satisfactorily if a great number of issues are to be given priority at the same time.

A further implication is that the corporation will be well advised to adopt an aggressive posture on social issues that bear closely on its business, those issues which have an important bearing on the success of the business over the long term. Why? Effective responsiveness may, of course, preempt regulation and mollify critics.

There are also two more important, less obvious reasons. The first is to preserve the freedom to define within the corporation what is acceptable behavior and what is unacceptable behavior within an ethical framework that sets the law as its lower boundary. This freedom carries with it the right to establish management incentives necessary to build an effective and purposeful organization.

A recent California case involving discrimination in a bank provides an apt illustration. The bank had been a bit backward in its thinking, to say the least, and a suit was brought against it by a coalition of human-rights groups. Finally, a consent decree was entered into which specified that 50% of those receiving promotions in that bank over the next 10 years had to be either women or minorities. Now imagine what that did to the seniority system in the bank and, indeed, to the chief executive's ability to manage his own organization.

Finally, strategic opportunities are likely to open up for a corporation which acts aggressively on social issues critical to its operations. There may be sources of competitive advantage in efforts to treat employees more equitably and humanely, to develop better safeguards for the environment, and so forth. Considerable money and energy are devoted to responding to social demands; the firm that can learn how to execute these responses more effectively and can manage the challenge of organizational learning is likely to be a more successful competitor in the long run.

Bull. N.Y. Acad. Med.
Vol. 54, No. 8 September 1978

Bibliography

Bibliography

Technical Articles

American Industrial Health Council, Inc. "AIHC Recommended Framework for Identifying Carcinogens and Regulating Them in Manufacturing Situations," New York: October 11, 1979, pp. B-1 to B-11.

Arthur, W.B. "The Economics of Risks to Life." *The American Economic Review,* March 1981, pp. 54-64.

Blot, W.J., Mason, T.J., Hoover, R., and Fraumeni, J.F., Jr. "Cancer by County: Etiologic Implications." *Origins of Human Cancer,* Cold Spring Harbor Laboratory, 1977, pp. 21-32.

Bohon, Robert L., M.D. "Industrial Considerations on the Proposed Testing and Good Laboratory Practices." *Toxic Control,* pp. 25-39, vol. iv, Edited by Marshall Lee Miller, Proceedings of the 4th Toxic Control Conference, Washington, D.C.: Government Institutes, Inc., December 10-11, 1979.

Brusick, David J. "Utility of Short-term Genetic Tests in Chemical Safety Evaluation; Evolving Issues." *Toxic Control,* pp. 40-46, vol. iv, Edited by Marshall Lee Miller, Proceedings of the 4th Toxic Control Conference, Washington, D.C.: Government Institutes, Inc., December 10-11, 1979.

Chelius, James R. "Economic and Demographic Aspects of the Occupational Injury Problem." *Quarterly Review of Economics and Business,* vol. 19, pp. 65-70.

Conference on Environmental Law—Toxic Substances, Proceedings. Marshall-Wythe School of Law, College of William and Mary, Williamsburg, Virginia: February 9-10, 1979.

Friess, Seymour L. "Contributions of Statistics to the Analysis of Environmental Health Problems Caused by Pollutants." *The American Statistician,* vol. 31, no. 1, American Statistical Association, February, 1977, pp. 2-7.

Gori, Gio Batta. "The Regulation of Carcinogenic Hazards." *Science,* vol. 208, April 18, 1980, pp. 256-261.

Johnson, Deborah G. "Body Integrity, Consent, and Toxic Substances." *Toxic Substances Journal,* vol. 1, no. 4, Spring, 1980, pp. 298-305.

Journal of the Washington Academy of Sciences. See Sections on "Carcinogens—Safe Doses?", pp. 62-90, "Air Pollutants—Safe Concentrations?", pp. 91-112, and "Occupational Exposures—Thresholds?", vol. 64, no. 2, Symposium Issue, Washington, D.C., June, 1974, pp. 126-155.

Kang, Han K. and Infante, Peter F. "Occupational Lead Exposure and Cancer." *Science,* vol. 207, February 29, 1980, pp. 935-36.

Karrh, Bruce W., M.D., and Pell, Sidney, Ph.D. "Brain Cancer in the Du Pont Company." Paper presented at the Workshop on "Brain Cancer in the Chemical Industry," New York Academy of Sciences, New York, October 27-29, 1980.

Kolata, Gina Bari. "Testing for Cancer Risk." *Science,* vol. 207, February 29, 1980, pp. 967-69.

Miller, Lowell E. "Chemicals Program of the Organization for Economic Cooperation and Development (OECD)." *Toxic Control,* vol. iv, Edited by Marshall Lee Miller, Proceedings of the 4th Toxic Control Conference, Washington, D.C.: Government Institutes, Inc., December 10-11, 1979, pp. 198-200.

O'Berg, Maureen T. "Epidemiologic Study of Workers Exposed to Acrylonitrile." Reprinted from *Journal of Occupational Medicine,* vol. 22, no. 4, April, 1980, pp. 245-52.

Pell, Sidney, Ph.D. "An Evaluation of a Hearing Conservation Program—A Five-Year Longitudinal Study." *American Industrial Hygiene Association Journal,* February, 1973, pp. 82-91.

Pell, Sidney, Ph.D. "Mortality of Workers Exposed to Chloroprene." Reprinted from *Journal of Occupational Medicine,* vol. 20, no. 1, January, 1978, pp. 21-29.

Pell, Sidney, Ph.D. "The Epidemiological Approach." *Environmental Health Perspectives,* vol. 26, October, 1978, pp. 269-273.

Pell, Sidney, Ph.D. "The Identification of Risk Factors in Employed Populations." Reprinted from *Transactions of the New York Academy of Sciences,* Series II, Volume 36, No. 4, April, 1974, pp. 341-356.

Pell, Sidney, Ph.D., O'Berg, Maureen T., and Karrh, Bruce W., M.D. "Cancer Epidemiologic Surveillance in the Du Pont Company." *Journal of Occupational Medicine,* vol. 20, no. 11, November, 1978, pp. 725-40.

Schneiderman, Marvin A., Ph.D. "Scientific Issues in a National Cancer Policy." *Toxic Control,* vol. iv, Edited by Marshall Lee Miller, Proceedings of the 4th Toxic Control Conference, Washington, D.C.: Government Institutes, Inc., December 10-11, 1979, pp. 95-107.

Smith, Jeffrey R. "Beryllium Report Disputed by Listed Author," *Science,* vol. 211, February 6, 1981, pp. 556-57.

Smith, Robert Stewart. "The Impact of OSHA Inspections on Manufacturing Injury Rates." *The Journal of Human Resources,* vol. 14, no. 2, 1978, pp. 145-170.

Stone, B.J., Blot, W.J., and Fraumeni, J.F., Jr. "Geographic Patterns of Industry in the United States, An Aid to the Study of Occupational Disease." *Journal of Occupational Medicine,* vol. 20, no. 7, July, 1978, pp. 472-477.

Strelow, Roger. "Corporate and Individual Responsibilities and Some Suggestions on Preventive Law." *Toxic Control,* vol. iv, Edited by Marshall Lee Miller, Proceedings of the 4th Toxic Control Conference, Washington, D.C.: Government Institutes, Inc., December 10-11, 1979, pp. 170-76.

Tabershaw, Irving R., M.D. "The Health of the Enterprise." *Journal of Occupational Medicine,* vol. 19, no. 8, August, 1977, pp. 523-526.

Travis, C.C. "Application of Simulation Models to the Control of Toxic Substances." *Toxic Control,* vol. iv, Edited by Marshall Lee Miller, Proceedings of the 4th Toxic Control Conference, Washington, D.C.: Government Institutes, Inc., December 10-11, 1979, pp. 73-86.

General Articles

AEI Journal on Government and Society. "Regulating Cancer—Fast, Fast, Fast Relief." *Regulation,* March-April, 1980, pp. 4-7.

AFL-CIO Executive Council. *OSHA: Years of Frustration.* Reprinted from *AFL-CIO American Federationist,* April-May, 1975.

Alsop, Ronald. "The Poison Brigade: Need for Toxicologists Soars as Firms Widen Product-Safety Tests." *Wall Street Journal,* vol. 53, no. 46, September 4, 1980, pp. 1, 20.

American Society for Personnel Administration. *Occupational Safety & Health Review.* Quarterly Report, March, 1977.

Ashford, Nicholas A. "Worker health and safety: an area of conflicts." *Monthly Labor Review,* September, 1975, pp. 3-11.

Atallah, Sami. "Assessing and managing industrial risk." *Chemical Engineering,* September 8, 1980, pp. 94-103.

Barnard, Robert C. "Identification of Toxic Substances Statutory Requirements." Paper presented at Toxic Substances Regulation Conference, November 8, 1979, Washington.

Barnard, Robert C. "The Emerging Regulatory Dilemma." Paper presented at New York Academy of Sciences, Workshop on The Management of Assessed Risk for Carcinogens, March 17, 1980, Prepublication Print.

Block, Duane L., M.D. "Organizational Environment and Ethical Conduct in Occupational Medicine: Loyalty of the Occupational Physician." *Bulletin of the New York Academy of Medicine,* vol. 54, no. 8, September, 1978, pp. 742-747.

Bureau of National Affairs, Inc. "NIOSH to Increase Efforts to Find Links Between Chemicals, Brain Tumors." *Daily Labor Report,* A-3, no. 211, October 29, 1980.

Bureau of National Affairs, Inc. "Occupational Health Efforts of Fifteen Major Labor Unions." Part 2 of *What's New in Collective Bargaining Negotiations and Contracts,* No. 818, October 7, 1976.

Bureau of National Affairs, Inc. "Report on Negotiated Safety Programs Compiled by Center for Labor Research and Education, University of California." *Daily Labor Report,* E-1—E-6, no. 219, November 13, 1973.

Bureau of National Affairs, Inc. "Selected Comments Relating to EEOC Proposed Guidelines on Employment Discrimination and Reproductive Hazards (Text)." *Daily Labor Report,* G-1—G-12, no. 133, July 9, 1980, Washington.

Bureau of National Affairs, Inc. "Statements Before House Labor Standards Subcommittee by ACTWU and Representative Butler Derrick (D-SC) on Byssinosis, or 'Brown Lung' (Text)." *Daily Labor Report,* E-1—E-6, no. 66, April 4, 1979.

Business Week. "A legal time bomb for corporations." June 16, 1980, pp. 150-154.

Business Week. "The new activism on job health." September 18, 1978, pp. 146-150.

Chemical Manufacturers Association. "Biomonitoring Helps Protect Environment." *Chemecology,* October, 1980, p. 8.

Chemical Manufacturers Association. "Workplace Safety: A Public Trust." *Chemecology,* October, 1980, p. 6.

Chemical Week. "Wanted: a sharper focus for research on cancer." September 24, 1980, p. 43.

Connolly, Walter B., Jr., and Crowell, Donald R., II. "Defending on OSHA Citation Case." *Employee Relations Law Journal,* vol. 2, no. 4, Spring 1977, pp. 465-480.

Crapnell, Stephen G. "Occupational health programs continue rapid growth." *Occupational Hazards,* June, 1980, pp. 49-54.

Crapnell, Stephen G. "What's causing cancer among auto workers?" *Occupational Hazards,* November, 1980, pp. 42-46.

Demopoulos, Harry, M.D. "Environmentally Induced Cancer . . . Separating Truth From Myth." Paper presented to Synthetic Organic Chemical Manufacturers Association, Inc., October 4, 1979, Hasbrouck Heights, N.J.

Dinman, Bertram D., M.D. "The Loyalty of the Occupational Physician." *Bulletin of the New York Academy of Medicine,* vol. 54, no. 8, September, 1978, pp. 729-732.

Du Pont. "Du Pont's Occupational Health and Safety Program." *Du Pont Management Bulletin,* vol. 9, no. 1, Wilmington, Del.: Public Affairs Dept., 1980, pp. 1-8.

Engel, Paul G. "How three majors pinpoint safety responsibility." *Occupational Hazards,* June, 1980, pp. 41-45.

Engel, Paul G. "Narrowing industrial hygiene's manpower gap: Are we making progress?" *Occupational Hazards,* March, 1980, pp. 69-72.

Engel, Paul G. "Respiratory protection: Cornerstone of this top-flight industrial hygiene program." *Occupational Hazards,* April, 1980, pp. 72-76.

Fleming, Richard. "Who Should Tell the Worker?" *Toxic Substances Journal,* vol. 2, no. 1, Summer 1980, pp. 25-34.

Harlow, Dan R., J.D., Ph.D. "Toxics Control Developments at OSHA, FDA and CPSC." *Toxic Control,* vol. iv, Edited by Marshall Lee Miller, Proceedings of the 4th Toxic Control Conference, Washington, D.C.; Government Institutes, Inc., December 10-11, 1979, pp. 132-37.

Hayes, Thomas C. "Complying With E.P.A. Rules." *The New York Times,* New York: January 16, 1980, pp. D1, D4.

Higginson, John. "Cancer and Environment: Higginson Speaks Out." *Science,* vol. 205, September, 1979, pp. 1363-1366.

Hilker, Robert, M.D. "If Hippocrates were Alive." *Bulletin of the New York Academy of Medicine,* vol. 54, no. 8, September, 1978, pp. 764-771.

Hoge, Warren. "New Menace in Brazil's 'Valley of Death' Strikes at Unborn." *The New York Times,* September 23, 1980.

Hudson, Richard L. "Hazardous Duty: Lack of Medical Staffs Heightens Health Peril at Some Smaller Plants." *Wall Street Journal,* April 20, 1979.

Imbus, Harold R., M.D. "Medical monitoring and surveillance in action." *Occupational Hazards,* September, 1980, pp. 40-44.

Industrial Health Foundation, Inc. *Industrial Hygiene Digest.* vol. 41, no. 9, September, 1977, Pittsburgh.

Industrial Union Department, AFL-CIO. "Employers Report Fewer Fatalities, Injury and Illness Reports Rise." *IUD Spotlight,* vol. 6, no. 4, 1977, pp. 1, 3.

Industrial Union Department, AFL-CIO. "Industry Tries to Muzzle Scientists." *IUD Spotlight,* vol. 7, no. 1, 1978, pp. 1, 8.

Industrial Union Department, AFL-CIO. "NIOSH Found Thousands of Workers with High Risk of Bladder Cancer—What Next?" *IUD Spotlight,* vol. 6, no. 4, 1977, pp. 6-7.

International Federation of Petroleum and Chemical Workers. *Environmental Control, and Health and Safety in the Workplace.* Panel Discussion, Prepared from the Congress records of a panel session during the Sixth World Congress of IFPCW, Istanbul, Turkey, June 15-19, 1970.

International Labour Organisation. *The Chemical Industries and the Working Environment.* Chemical Industries Committee, Report II, Eighth Session, Geneva: International Labour Office, 1976.

Karrh, Bruce W., M.D. "A Company's Duty to Report Health Hazards." *Bulletin of the New York Academy of Medicine,* 2nd series, vol. 54, no. 8, September, 1978, pp. 782-88.

Karrh, Bruce W., M.D. "Medical Monitoring." *National Safety News,* August, 1976, pp. 68-70.

Karrh, Bruce W., M.D. "The Confidentiality of Occupational Medical Data." *Journal of Occupational Medicine,* vol. 21, no. 3, March, 1979, pp. 157-160.

Lehmann, Phyllis. "Cancer—one in four will get it." *Hazard,* vol. 1, no. 1, (January, 1976), Reprint *Job Safety and Health,* U.S. Dept. of Labor, July, 1976, pp. 4-6.

Lieberman, Marvin, J.D., Ph.D. "Summary of the Discussion: The Duty to Report Hazards." *Bulletin of the New York Academy of Medicine,* vol. 54, no. 8, September, 1978, pp. 795-796.

Lieberman, Marvin, J.D., Ph.D. "Summary of the Discussion: The Issue of Confidentiality." *Bulletin of the New York Academy of Medicine,* vol. 54, no. 8, September, 1978, pp. 772-773.

Lieberman, Marvin, J.D., Ph.D. "Summary of the Discussion: The Organizational Environment and Ethical Conduct in Occupational Medicine." *Bulletin of the New York Academy of Medicine,* vol. 54, no. 8, September, 1978, pp. 724-728, and pp. 748-750.

Martin, Donald L., Ph.D. "The Organizational Environment and Ethical Conduct in Occupational Medicine." *Bulletin of the New York Academy of Medicine,* vol. 54, no. 8, September, 1978, pp. 715-719.

McLean, Alan, M.D. "The Issue of Confidentiality." *Bulletin of the New York Academy of Medicine,* vol. 54, no. 8, September, 1978, pp. 751-757.

McNeil, Donald G., Jr. "Dumped Hooker Pesticides Poisoned Wells on Coast." *The New York Times,* August 6, 1979.

McNeil, Donald G., Jr. "Hooker Corporation Papers Indicate Management Sanctioned Polluting." *The New York Times,* August 5, 1979, pp. 1, 39.

McNeil, Donald G., Jr. "House Panel Denied Data on Water Contamination." *The New York Times,* May 16, 1979, p. A-12.

McNeil, Donald G., Jr. "Two Studies Compete on Peril at Hooker." *The New York Times* (Special), April 23, 1979.

Miller, Judith. "S.E.C. Says Occidental Hid Potential Liabilities." *The New York Times,* July 3, 1980, pp. D1, D6.

Mosher, Lawrence. "Big Steel Says It Can't Afford to Make the Nation's Air Pure." *National Journal,* July 5, 1980, pp. 1088-1092.

Mosher, Lawrence. "Love Canals by the Thousands—Who Should Pay the Costly Bill?" *National Journal,* May 24, 1980, pp. 855-57.

Neal, Alfred C., Ph.D. "The Organizational Environment and Ethical Conduct in Occupational Medicine." *Bulletin of the New York Academy of Medicine,* vol. 54, no. 8, September, 1978, pp. 720-723.

Occupational Hazards. "In 1980, NIOSH will step up field investigations." December, 1979, pp. 38-39.

Occupational Hazards. "Occupational disease—Casualties: 2 million workers; Price-tag: $11.4 billion." April, 1980, pp. 95-99.

Occupational Hazards. "Toxicology—much in demand, much misunderstood." June, 1980, pp. 54-55.

Occupational Hazards. "Were Air Samples Valid?" January, 1980, pp. 49-50.

Palisano, Peg. "The war on cancer: no quick victory in sight." *Occupational Hazards,* April, 1980, pp. 87-91.

Palisano, Peg. "1981-85—These occupational health issues will be in the forefront." *Occupational Hazards,* October, 1980, pp. 79-82.

Pappert, Muriel S., R.N., M.A. "The Issue of Confidentiality: A Perspective from the Nursing Profession." *Bulletin of the New York Academy of Medicine,* vol. 54, no. 8, September, 1978, pp. 758-763.

Reutter, Mark. "Workmen's Compensation Doesn't Work or Compensate." *Business and Society Review,* No. 35, Fall, 1980, pp. 39-44.

Rivera, Anne. "Union Action Solves Eye Burn Mystery," *ACTWU Labor Unity,* Amalgamated Clothing and Textile Workers Union, September, 1980, pp. 11-12.

Rogers, James. "Hazardous Waste Regs Increase in Complexity." *Legal Times of Washington,* February 9, 1981, pp. 17-19, 28-9, 32.

Rothmyer, Karen. "Brown Lung's Legacy." *The Wall Street Journal,* June 7, 1976, p. 26.

Sand, Robert H. "Current Developments in OSHA: Access of NIOSH To Employee Medical Records." *Employee Relations Law Journal,* vol. 6, no. 2, pp. 304-308.

Sand, Robert H. "Current Developments in OSHA: Regulations, Proposed Regulations, and Interpretive Bulletins." *Employee Relations Law Journal,* vol. 6, no. 1, pp. 145-150.

Scala, Robert A., Ph.D. "The Duty to Report Hazards: A Toxicologist's View." *Bulletin of the New York Academy of Medicine,* vol. 54, no. 8, September, 1978, pp. 774-781.

Schept, Kenneth. "Asbestos Suits Catching Fire." *National Law Journal,* vol. 2, no. 49, August 18, 1980, pp. 1, 10.

Schuman, Bernard J., M.D. "Physicians and Patients in the Occupational Setting: The Rules of the Game." Commentary Reprinted from the *Journal of the American Medical Association,* vol. 244, no. 21, November 28, 1980, pp. 2417-18.

Severo, Richard. "Federal Mandate for Gene Tests Disturbs U.S. Job Safety Official." *The New York Times,* February 6, 1980, pp. A1, A17.

Severo, Richard. "Genetic Tests by Industry Raise Questions on Rights of Workers." *The New York Times,* February 3, 1980, pp. 1, 36.

Severo, Richard. "Screening of Blacks by Du Pont Sharpens Debate on Gene Tests." *The New York Times,* February 4, 1980, pp. A1, A13.

Singer, James W. "Should Equal Opportunity for Women Apply to Toxic Chemical Exposure?", *National Journal,* October 18, 1980, pp. 1753-55.

Shabecoff, Philip. "Industry and Women Clash over Hazards in Workplace." *The New York Times,* January 3, 1981.

Sheridan, Peter J. "What's causing mysterious illnesses?—NIOSH seeks answers." *Occupational Hazards,* April, 1980, pp. 63-69.

Shriver, Donald W., Jr., Ph.D. "Legal and Philosophical Perspectives on Ethical Issues: A Theologian's View." *Bulletin of the New York Academy of Medicine,* vol. 54, no. 8, September, 1978, pp. 797-807.

Smith, R. Jeffrey. "Government Says Cancer Rate Is Increasing." *Science,* vol. 209, August 29, 1980, pp. 998-1002.

Stansbury, Jeff. "Tracking Down the Time-Bomb Killers: UAW Opens a New War on Cancer in the Workplace." *Solidarity,* vol. 23, no. 6, May 16-31, 1980, pp. 10-14.

Tabershaw, Irving R., M.D. "Summary." *Bulletin of the New York Academy of Medicine,* Conference on Ethical Issues in Occupational Medicine, vol. 54, no. 8, September, 1978, pp. 810-817.

Van Liere, Kent D. and Dunlap, Riley E. "The Social Bases of Environmental Concern: A Review of Hypotheses, Explanations and Empirical Evidence." *Public Opinion Quarterly,* The Trustees of Columbia University, 1980, pp. 181-197.

Wegman, Donald H., M.D. "Duty to Report Hazards: A Public-Health Perspective." *Bulletin of the New York Academy of Medicine,* vol. 54, no. 8, September, 1978, pp. 789-794.

Wessel, Milton R., Esq. "Medical Ethics in Litigation." *Bulletin of the New York Academy of Medicine,* vol. 54, no. 8, September, 1978, pp. 808-809.

Whorton, Donald, M.D. and Davis, Morris E., J.D. "Ethical Conduct and the Occupational Physician." *Bulletin of the New York Academy of Medicine,* vol. 54, no. 8, September, 1978, pp. 733-741.

Williams, Senator Harrison A., Jr. "OSHA Under Inspection." *Trial Magazine,* vol. 11, no. 5, September-October 1975, pp. 12, 20, 30.

Wood, Norman J. "Environmental Law and Occupational Health." *Labor Law Journal,* March, 1976, pp. 152-162.

Zener, Robert V., and Olstein, Myron. "Pollution Regs: Problems for Lawyers, Accountants." *Legal Times of Washington,* September 22, 1980, pp. 10-11.

Zuesse, Eric. "Love Canal: The Truth Seeps Out." *Reason,* vol. 12, no. 10, Santa Barbara, California: February, 1981, pp. 16-33.

Books

Amalgamated Clothing and Textile Workers Union, AFL-CIO, CLC; *The Right to Breathe: Where did it go and how did I lose it?*; December, 1980.

Ashford, Nicholas A. *Crisis in the Workplace: Occupational disease and injury, a report to the Ford Foundation.* Cambridge, Mass.: MIT Press, 1976.

Barth, Peter S., with Hunt, H. Allan. *Workers' Compensation and Work-related Illnesses and Diseases.* Cambridge, Mass.: MIT Press, 1980.

Chemical Manufacturers Association. *Worker Safety in the Chemical Industry—What We're Doing About It.* Washington: May, 1980.

Chown, Paul. *Workplace Health and Safety: A Guide to Collective Bargaining.* Regents of the University of California, Berkeley, 1980.

Dominguez, George S., *The Business Guide to Tosca: Effects and Actions.* New York: John Wiley & Sons, Inc., 1979.

Egdahl, Richard H., ed. *Background Papers on Industry's Changing Role in Health Care Delivery.* Springer Series on Industry and Health Care, no. 3, New York: Springer-Verlag, 1977.

Egdahl, Richard H. and Walsh, Diana Chapman, M.S., eds. *Health Services and Health Hazards: The Employee's Need to Know.* Industry and Health Care, no. 4, New York: Springer-Verlag, 1978.

Egdahl, Richard H., M.D. and Walsh, Diana Chapman, M.S. *Sounding Board: Industry-Sponsored Health Programs: Basis for a New Hybrid Prepaid Plan.* Reprinted from the *New England Journal of Medicine,* 296:1350-1353 (June 9,) 1977, Massachusetts Medical Society.

Epstein, Samuel S., M.D. *The Politics of Cancer.* New York: Anchor Press/Doubleday, 1979.

Kochan, Thomas A., Dyer, Lee, and Lipsky, David B. *The Effectiveness of Union-Management Safety and Health Committees.* Michigan: W.E. Upjohn Institute for Employment Research, 1977.

Marsh & McLennan Companies, Inc. *Risk in A Complex Society: A Marsh & McLennan Public Opinion.Survey.* Conducted by Louis Harris and Associates, Inc., 1980.

O'Reilly, James T. *Unions' Rights to Company Information.* Labor Relations and Public Policy Series, No. 21, University of Pennsylvania: The Wharton School, 1980.

Silverstein, Michael, M.D. *The Case of the Workplace Killers: A Manual for Cancer Detectives on the Job.* Washington: International Union, UAW, November, 1980.

Government Publications

Federal Register. "Interagency Working Group on a Hazardous Substances Export Policy: Draft Report," vol. 45, no. 157, Tuesday, August 12, 1980, pp. 53754-53769.

National Institutes of Health, U.S. Department of Health, Education, and Welfare. "Cancer Mortality in U.S. Counties with Petroleum Industries." *Science,* vol. 198, October 7, 1977, pp. 51-53.

Toxic Substances Strategy Committee, Council on Environmental Quality. *Toxic Chemicals and Public Protection.* Report to the President, Washington: U.S. Government Printing Office, May, 1980.

U.S. Department of Health and Human Services, Public Health Service. *Environmental Health, a plan for collecting and coordinating statistical and epidemiologic data.* Washington, 1980.

U.S. Department of Labor, Bureau of Labor Statistics. *Occupational Injuries and Illnesses in the United States by Industry, 1978.* Bulletin 2078, (U.S. Government Printing Office, Washington, D.C.), August, 1980.

U.S. Department of Labor, Bureau of Labor Statistics. *Occupational Safety and Health Statistics: Concepts and Methods,* Report 438, 1975.

U.S. Department of Labor, Bureau of Labor Statistics, *What Every Employer Needs to Know About OSHA Recordkeeping,* Report 412-2, November, 1978.

U.S. Department of Labor, Occupational Safety and Health Review Commission, and U.S. Department of Health, Education, and Welfare. *The President's Report on Occupational Safety and Health,* Washington, 1975.

Related Conference Board Reports

Report No. 811